A HANDBOOK ON THEORY AND PRACTICE OF

ILLUMINATING ENGINEERING

A HANDBOOK ON THEORY AND PRACTICE OF

ILLUMINATING ENGINEERING

VOLUME–ONE

[INCLUDING MANUFACTURING PROCESS OF GLS, AUTO, MINI AND HALOGEN LAMPS]

By

KAMALESH ROY

AMBIM (Lond.), PGIM (Glasgow), AMIMI (Lond.)
Industrial Advisor to
S.I.S.I. Bihar and U.P.
For Electric Lamp Manufacturing Industries

UNIVERSITY SCIENCE PRESS

(An Imprint of Laxmi Publications Pvt. Ltd.)

BANGALORE • CHENNAI • COCHIN • GUWAHATI • HYDERABAD
JALANDHAR • KOLKATA • LUCKNOW • MUMBAI • RANCHI
NEW DELHI • BOSTON, USA

Printed and bound in India

First Edition : 2023; *Edition* : 2025

ISBN **:** 978-93-80386-55-3

Branches

✆	Bengaluru	080-26 75 69 30
✆	Chennai	044-24 34 47 26
✆	Guwahati	0361-254 36 69
✆	Hyderabad	040-27 55 53 83
✆	Kochi	0484-405 13 03
✆	Kolkata	033-40 04 77 79
✆	Lucknow	0522-430 36 13
✆	Ranchi	0651-224 24 64

Published in India by

UNIVERSITY SCIENCE PRESS

(An Imprint of Laxmi Publications Pvt. Ltd.)

An ISO 9001:2015 Company

113, GOLDEN HOUSE, GURUDWARA ROAD, DARYAGANJ,

NEW DELHI - 110002, INDIA

Corporate Office : B-33, Sector-5, Noida,

Gautambuddha Nagar (U.P.) - 201 301, INDIA

Telephone : [illegible]

www.laxmipublications.com

info@laxmipublications.com

Preface

Electric Lamp Manufacturing Industry in India has proudly celebrated its Golden Jubilee Year in 1932. In 1987, India will be producing 400 million GLS Lamps for the home market and also export millions of auto lamps and mini lamps to all the neighbouring countries. It's a pulsating, promising, profiteering and prestigeous industry of India.

Even in the advanced countries, Illuminating Engineering is a fairly new stream in engineering studies. Naturally it might take a little more time to introduce Illuminating Engineering in our curriculum. The greatest set-back for its non-inclusion in the engineering course is non-availability of sufficient books on this subject. This was also the reason for its delayed introduction in the engineering stream in the advanced countries.

The present book is written to cover the first year course in Illuminating Engineering. The final task of an Illuminating Engineer is to provide aesthetic presentation of any field of vision. At the beginning he has to make different types of lamps to light up the field and the tungsten-filament lamp is still his best choice.

Throughout the book, efforts have been made to bridge the gap between 'theory of lamp making' and the 'technology of manufacturing lamp'. The older generation on the shop floor have a natural tendency to scoff at the theory, so, the early chapters have been made as easy as possible which at places again may prove to be too easy for the advanced engineering-students. For the present, this volume may suffice also as the syllabus of a "Short-Diploma Course on Tungsten Filament Lamp Making" under Electrical Engineering.

To complete the course on Illuminating Engineering, it is hoped the volumes on "Gas Discharge Lamp" and "Principles and Applications of Illumination" will follow this publication.

Manufacturing process of electric lamps varies from manufacturer to manufacturer. Then again, there are large scale, small scale and tiny scale lamp factories where they adopt their own ways and means for the technology of manufacture. But one thing is common to all—that is to make finally a good lamp. So, the Lamp Makers are requested to select and follow the process as suitable to them for making good lamps.

Most of the writings are taken from my lecture notes to the Lamp Makers. To collect them from various places, scrutinise and elaborate them and finally take the decision to publish them in a book form possibly could never be done by me alone. I am grateful to Mr. K.K. Manchanda, Deputy Director, Small Industries Service Institute who had been the source of inspiration to start this venture.

I am thankful to Mr. B.L. Narula, Mr. A.K. Mondal, all the kindly staff and Librarian of Indian Institute of Technology, Kanpur for their constant help to complete this book.

I also thank Mr. V.P. Subrahmanian for his patience and zeal to type and retype the manuscript from many corrections and scribblings. I also record my adoration and love for the three lights of my life Sudip, Nova, Bhaskar and my dear wife Lakshmi for their continued inspiration and support by bearing with me through these four years of long struggle to write this book.

I am most grateful to Prof. Sampath, Director, I.I.T., Kanpur for his constant encouragement while writing this book and giving valuable suggestions for introducing Illuminating Engineering in our curriculum. I record my heartfelt gratitude for his kind and encouraging "Foreword" of this book.

Most probably this being the very first book in Illuminating Engineering to be published from India, the response from our electric lamp industries to supply some informations have been very disheartening. I expect for the next two books our industries would contribute sufficient informations so the world may know about the progress we have been making in the field of Illuminating Engineering.

I am most grateful to Mrs. C.J.M. Daalder-de-Leeuw of Philips Technical Library, Holland for her untiring and continuous supply of informations to my every query.

I record my sincere indebtedness for the help and guidance received from Mr. T.V. Mansukhani, Mg. Director, H.M.T. Ltd.; Dr. George Wise & Mr. Gilbert H. Reiling of General Electric Company, Ohio, USA; Dr. Avinash Kulkarni, Mg. Director, Litex Elect., Ltd., Dr. K.S. Sharma Head of Optics Div., National Physical Laboratory; Director, Indian Standards Institute; The Managing Director of Badalex, a Division of Sale Tinley Technology Plc., England and Dr. Hira Sadhak, the renowned Economic Advisor.

Suggestions from academicians for improving on the course of Illuminating Engineering and from Lamp Technologists for special features of lamp making are most sincerely invited.

I shall feel myself greatly rewarded if our Lighting Engineers, Lamp Technologists and the new entrepreneurs in the lamp industry try to follow the modern trends based on scientific research rather than continuing with the old procedures.

To the Illuminating Engineers of advanced countries, I may humbly point out that the lamps made in India also follow the EEC Norms and give very high quality performance though some of these are made in very slow moving and crude machines. The time is not far when some inventions of our lamp technologists will be adopted the world over. In fact our Indian Illuminating Engineers have won many International Awards on illumination which is proud testimony of our determination, zeal and innovative qualities to be on the fore-front in the field of illumination.

It needs great courage and sense of responsibility to the society for a publisher to takeup the works of such a maiden venture. Mr. R.K. Gupta, B.Sc. Engg. (Mech.) Managing Director, Laxmi Publications, has shown these rare qualities in publishing this present volume for which he deserves my everlasting gratitude.

1-8-1987

Aalo Bhavan
111-A/352, Ashok Nagar
Kanpur-208012

–KAMALESH ROY

Acknowledgments

The author thankfully acknowledges the contributions from the following organisations:

1. Philips Technical Library, Holland.
2. General Electric Company, Cleveland, Ohio, U.S.A.
3. Badalex—a Division of Sale Tinley Technology Plc., England.
4. Hindustan Machine Tools Ltd., Hyderabad, India.
5. Litex Electricals Pvt. Ltd., Pune.
6. Philips Professional Journal, Bombay.
7. Indian Standards Institute, New Delhi.
8. National Physical Laboratory, New Delhi.
9. Sudip Electrical Works, Calcutta.
10. Kerala Lamp Mfg. Co., Taliparamba, Kerala.
11. Crafts de Fluorescent, Bombay.
12. Automobile Bulb Co., Kanpur.
13. Kalpaka Lamp Works, Taliparamba, Kerala.
14. Navjyoti Lamp Works, Patna.
15. Vishal Auto Lamps, Kanpur.

Contents

PART I

THEORY

PART II

PRACTICE

With love, respect and gratitude

Dedicated to my

Father Late Mr. K.M. Roy, A.G.W.I.

and

Mother Mrs. Renuka Roy

who brought for me the

very first ray of

LIGHT

ILLUMINATING ENGINEERING

PART I
THEORY

"We all know what light is;
but it is not easy to say
what it is."

—Samuel Johnson.

and the Lord Said

"LET THERE BE LIGHT" ....*Book of Genesis*

....*and God Created Light.*

Of all the five organs of perception, the eyes which perceive light, play the most vital role in understanding the wonders of this creation and inducing the human beings to participate in the advancement of the civilization. It's through this medium of light that we receive ninety per cent information of the world and one tenth of our cerebral cortex is engaged to interpret the datas collected by the light alone. So, in the absence of sufficient natural light or in the darkness of night, the human eyes become irresponsive and all activities practically stop. Production of 'Artificial Light' then becomes indispensible for the working process to continue............and we manufacture lamps.

1

Natural Light

The principal aim of *'artificial lighting'* is to help in the best functioning of the human vision. This preimposes the condition that artificially produced light should have the characteristics as near as possible to that of *'natural light'* which stimulate the optic nerves.

To produce an *'incandescent electric lamp'* offering such service in quality and quantity of light at every demanding situation and at the same time also to be within the economic reach of users, 'The lamp maker' need the support of many branches of 'science and technology', particularly the chapters on 'Light, Metals and Chemicals, Electricity, Glass Making, Pressure and Vacuum Operations'.

What is Light ?

To produce light artificially, some of the essential characteristics of natural light are to be noted earlier. Alike all other energies as heat, sound, electricity etc. light energy also by itself is invisible but it's presence make other objects visible. Light travels from one body to another without any help from any materials in the intervening space. It travels in a straight line at a constant speed through the same medium and would change it's course and speed when confronted with a different medium.

All these may depict some of the behaviours of light but cannot explain it's physical shape and entity. Many theories have been advanced to explain 'what light is', which only have kept it beyond common man's easy understanding and imagination.

The 'wave theory' of light was propounded by Huygens in 1678 which ran parallel with 'corpuscular theory' of Newton. In 1675, Römer, the Danish astronomer discovered the velocity of light about 3×10^{10} cms per second. In 1850, Foucaults proved that the velocity of light in denser medium is less than the lighter medium or rarer medium.

The 'electromagnetic wave theory' of light was first propounded by Clark Maxwell in 1873. While working with displacement of currents in dielectrics, Maxwell concluded that the electromagnetic disturbances travelled in space (v) following the equation

$$v = \frac{c}{\sqrt{\mu k}}$$

where μ = magnetic permeability

k = specific inductive capacity of the medium

c = ratio between electromagnetic and electrostatic quantity of electricity.

Maxwell based his theory on the findings of Weber and Kohlroush who had very carefully and painstakingly found out the value of 'c', the ratio of the units between electromagnetic and electrostatic quantity to be 3.107×10^{10} cms per second. As this velocity comes closer to the propagation of light, Maxwell concluded that light also was an electromagnetic wave. No doubt, the 'electromagnetic wave theory' of light successfully describes almost all the phenomenon of light but the 'Newtonian Corpuscular Theory' of light also holds an equally honoured position in optics and wave mechanics as we shall see in the next chapter.

It is easier to describe light as a form of radiant energy. For "Illuminating Engineering", the behaviour of light described in Electromagnetic Wave Theory as a form of radiant energy has been found suitable for all practical purposes. At particular wave length, with definite frequency, these electromagnetic waves can stimulate the receptors of eyes and make things visible to us.

Light as electromagnetic wave :

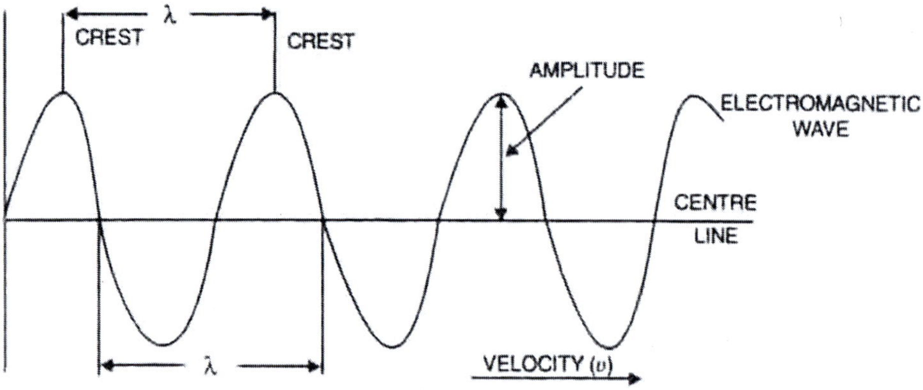

Fig. 1.1

The above illustration (Fig. 1.1) represents an electromagnetic wave. The following properties of an electromagnetic wave are also found in a light wave.

Wavelength. The distance starting from a point at the centre line to the next crossing point after one up and down movement is known as one unit of wavelength. It can also be measured from one crest point to the next crest point. The wavelength is represented by the letter λ (Lambda).

Frequency. The frequency of a chain of waves is defined as the number of waves passing through a given point in one second. This is denoted by n.

Velocity. Electromagnetic waves travel at a definite speed and is not dependent on the wavelength. Velocity of electromagnetic radiation or velocity of light is 3×10^8 metre per second. Velocity is represented by (υ) and the distance travelled by it as (S).

Amplitude. This is the height of the wave which shows the energy it contains. The energy is proportional to the square of the amplitude. Whereas the amplitude expresses the strength of the wave, the wavelength speaks of the nature of wave.

Some examples. When an Electric Generator runs, it also produces, electromagnetic waves. These are very long waves, extending from 1500 km. to 12,500 km. and their frequency is hardly 10 to 100 times in one second.

The frequency of a radio transmitter as 900 kilo cycles might sound very high and impressive but we never imagine that our own eyes are receiving a shocking hit of 500,000,000,000,000 times of light wave per second which make us see.

From above propositions it may be concluded now that

$$\text{Velocity} = \text{Frequency} \times \text{Wavelength}$$

$$v = n \times \lambda$$

$$v = n\lambda$$

velocity of light or electromagnetic propogation is a constant

$$= 3 \times 10^8 \text{ m per second}$$

So $$3 \times 10^8 = n \times \lambda$$

or $$\lambda = \frac{3 \times 10^8}{n}$$

This is an important formula which is often utilised to find either wavelength or frequency, when one is known.

Unit of Measurements. For measuring the above characteristics of light we have to select a very very small unit. For finer measurement in general engineering $\frac{1}{10,000}$ of a centimetre known as Micron is used.

$$1 \text{ Micron} = 10^{-4} \text{ cm}$$

$$1 \text{ Micron} = 25.4 \text{ Thou.}$$ (one Thou. is one thousandth part of an inch)

For measuring light the universal unit chosen is Angstrom (Å)

1 Angstrom = One hundred millioneth of a centimetre

$$= \frac{1}{100,000,000} \quad \text{or} = 10^{-8} \text{ cm}$$

Sometimes light is also measured in Nanometer

1 Nanometer = 1 Millimicron = One ten millioneth of a centimetre

$$= \frac{1}{10,000,000} \text{ cm} = 10^{-7} \text{ cm}$$

So from above we see that a Nanometer = 10 Å (Angstrom) unit. For understanding 1 Å (One Angstrom unit) it is to be remembered that this is the length of light in the X-ray' range and is invisible to human eye (Fig. 1.2). The longer the Angstrom unit, the nearer it comes to human visible spectrum. The electromagnetic waves those are visible to our eye range between 3800 Å to 7500 Å. When the length of light is less than one Angstrom and smaller, it becomes gamma ray and cosmic ray.

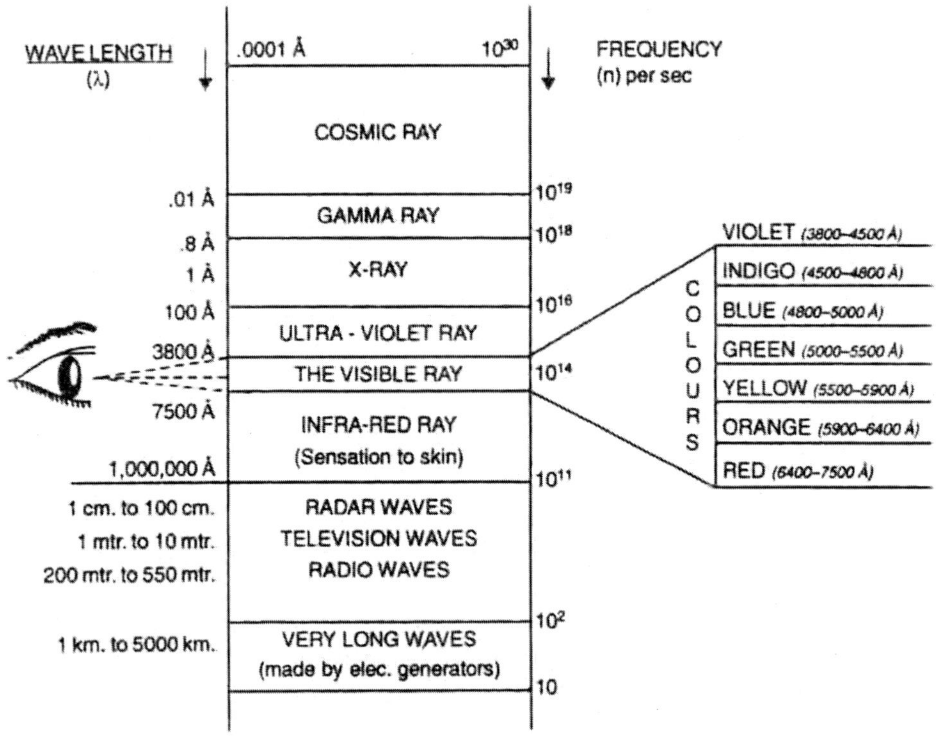

Fig. 1.2

A Simple Camera Compared with the Human Eye

The human eye is the organ to receive and response to the characteristics of illumination. The construction of the eye has guided investigators to designs and produce many types of lamps, colour effects and testing equipments.

The eye is primarily an electrochemical device. The radiant energy falling on the eye causes a chemical reaction which in turn produces electrical impulses and informs the brain to make decisions.

Unit of measurement for temperature :

In our daily work we use centigrade scale in which the freezing point of water is 0°C and the boiling point of water is 100°C at sea level.

For all scientific work, the temperature is recorded in Kelvin (K) scale. Kelvin scale is the absolute scale of temperature. Each degree of Kelvin (K) scale and centigrade (°C) scale is equal in value. The difference being that the zero of Kelvin scale is –273°C. So, the centigrade reading is to be added with 273° to convert it to Kelvin.

For example–

The colour temperatures of

1. Vacuum type tungsten filament is 2127°C or 2400°K.
2. Gas filling type tungsten filament is 2577°C or 2850°K.

(In this book for day to day work, the temperatures have been mentioned in centigrade and for theoretical/laboratory results Kelvin scale has been used).

To understand the construction of the eye a simple camera (Fig. 1.3) with shutter (Fig. 1.4) is explained here.

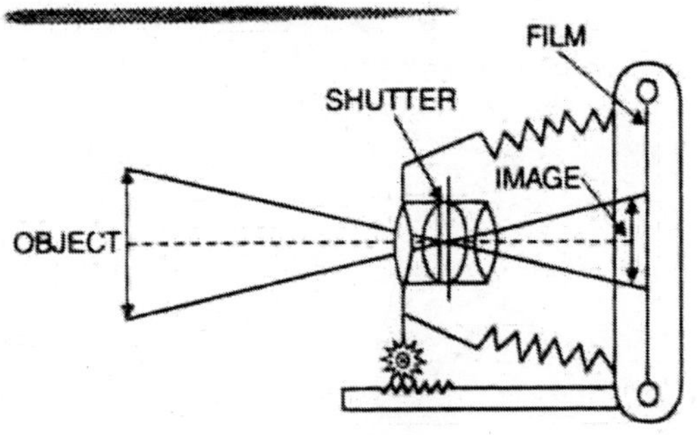

Fig. 1.3

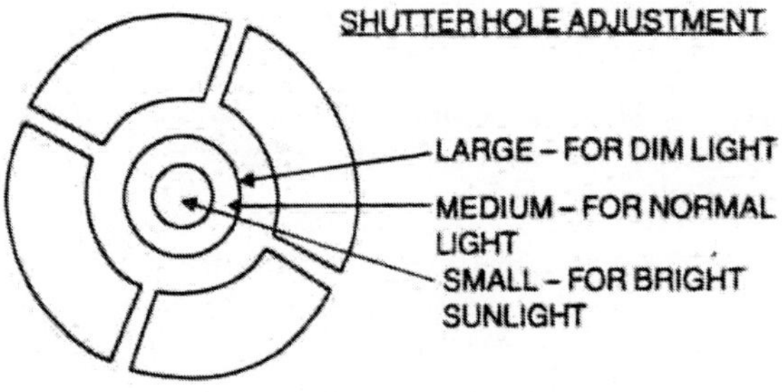

Fig. 1.4

In a Simple Camera

A shutter mechanism is adjusted to introduce more light falling on dim objects and to receive less light for bright objects.

In an Eye

1. The shutter represents the IRIS.
2. The lense is called the LENSE.
3. The film represents the RETINA.

LIGHT CENTRE APPARATUS

A simple locally made camera with a convex lense can be easily constructed to serve as a "Light Centre Apparatus".

In front of the camera a lamp holder is fixed on an adjustable stand. The rear of the camera has a projected end to avoid direct falling of light on to the screen. The screen is made of an oil paper which will show only the glowing filament. On this screen, the different heights of the lamp tube and stretching of the filament are marked for checking.

THE EYE BALL

The light enters into the eye ball through the cornea and passes through a 'water like substance' called AQUEOUS HUMOR.

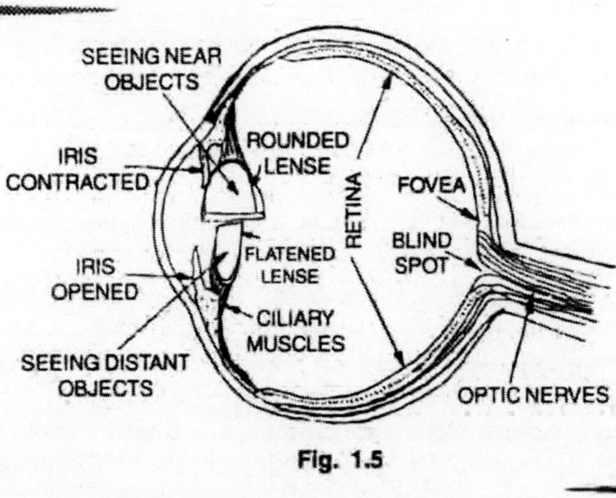

Fig. 1.5

As told earlier the IRIS is the shutter—a diaphragm which controls the amount of light by contracting or expanding. This comes into operation giving more opening for seeing dim objects and less opening when the object is bright.

The light then passes through the LENSE which is controlled by the CILIARY MUSCLES. As the lense has to produce always a sharp image on to the RETINA—whether the object is near or far; the shape of the lense is rounded for the near object and flattened for the distant object by the action of the ciliary muscles.

The light, thus controlled, then produces an inverted image on the RETINA or the 'recording film' as inside a camera. The retina covers almost two third of the inner surface of the eye. The sharpest point of the focus of the lense is called FOVEA. The retina consists of RODS and CONES (*Photo 1.1*) which are light sensitive receptors. These are mostly concentrated at the FOVEA, the final focussing point at the visual axis of the eye. The rods contain the photo chemical pigment named 'Rhodopsin' (visual purple) which undergoes chemical change soon light falls on it. The cones generally contain three types of visual pigments for colour vision. (*See Photo 1.2*)

Light is absorbed by the retinal pigments which undergo chemical change and in turn produces electrical potential differences across the retina. By means of the optic nerves (*Photo 1.2*) these electrical impulses are transmitted to the visual cortex of the brain—which takes about 2 milli seconds. Here the strength of the stimulus plays a very important part. There is a small gap of time between stimulation of the eye and the resulting sensation in the eye. In dim light the stimulation is less and the sensation created becomes slow. In continuous use the sensation period is increased and the eye has to work more and physical fatigue sets in.

The eye seems to be a simple superficial organ but in fact it is a highly complex, extensively anatomical and psychophysiological system. The eye is the sole judge of the artificially produced light. The complex system of working of the eye; it's behaviour in different environment, at different age levels and finally the psychological bearing of the individual—all are to be studied and considered while manufacturing lamps and creating artificial illumination.

When the eye is conditioned to dim light, the retinal rod receptors come into operation and the eye is known as dark-adapted. When the retinal cone receptors start operating the eye, it is called light-adapted. If a dark-adopted eye is suddenly exposed to bright light it will take about one hour for the eye to adapt itself to such condition. This is a very important point for the light manufacturers when they measure the brightness of lamps.

The eye is like an automatically operated switch or a gate valve which starts and stops functioning when radiation of particular wave lengths are fed into it. The following curve gives a clear picture how natural and mathematically correct is the function of the radiant energy on the optic nerves. Eye has the feeling of a cent per cent perception of luminosity of the centre band of the visible spectrum between green and yellow.

This is the reason why objects look less bright in red and violet light which are at the base ends of the curve. To judge and to explain the quality of light is a very difficult task as it will vary from person to person. Even with normal rod and cone vision of a set of healthy eyed individuals, there will be differences in the perception of spectral luminosity. Even for setting very sensitive automatic instruments for measuring light a first setting or comparison or standard has to be determined. In 1924 the International commission on illumination

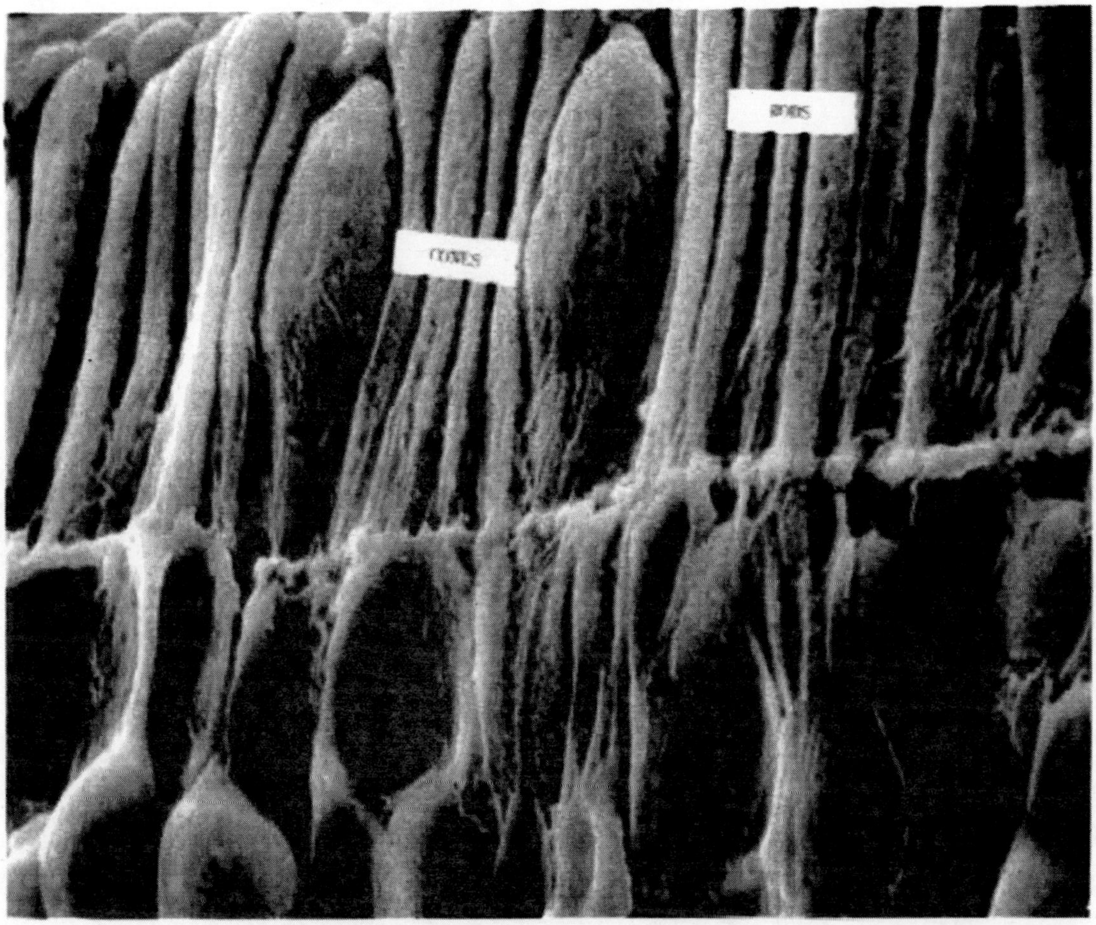

Photo 1.1. Microphotograph of rods and cones.

The retina of the eye is actually a part of the brain itself—the only point at which it receives stimuli directly from outside the body. Light shining through the eye's pupil and lens strikes the retina, where sensitive cells called rods and cones transform light energy into the brain's electrochemical language. The inner segments of some of these receptors, magnified 4,500 times, are shown above. The slender rods, which are more sensitive in dim light, report a simple black-and-white world, while the thicker cones, which need stronger stimulation, can distinguish colour. In all, there are some 120 million rods and seven million cones in each retina.

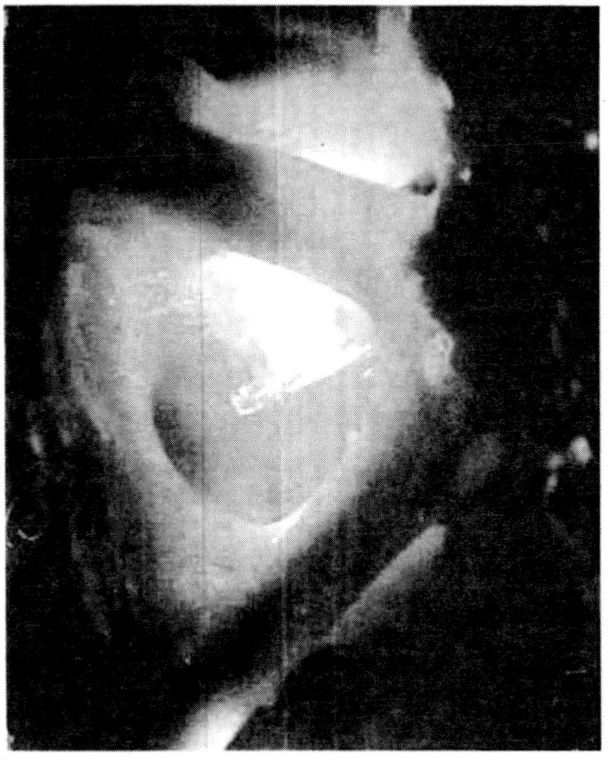

Photo 1.2. Two massive optic-nerve tracts, each carrying fibers from both eyes, branch out toward the brain's two hemispheres from the optic chiasma, a junction just behind the eyes. Each of the millions of fibers carries information from a single cell in the eye to its counterpart in the brain.

(Courtesy : Lennart Nilsson.)

collected a standard set of luminosity perception data of 200 observers and found that the light produces highest luminosity at 5,560 Å. We know this to be the length of the wave but we have to find out the frequency or the times of repetition of this length to hit our eyes which cause this brightness.

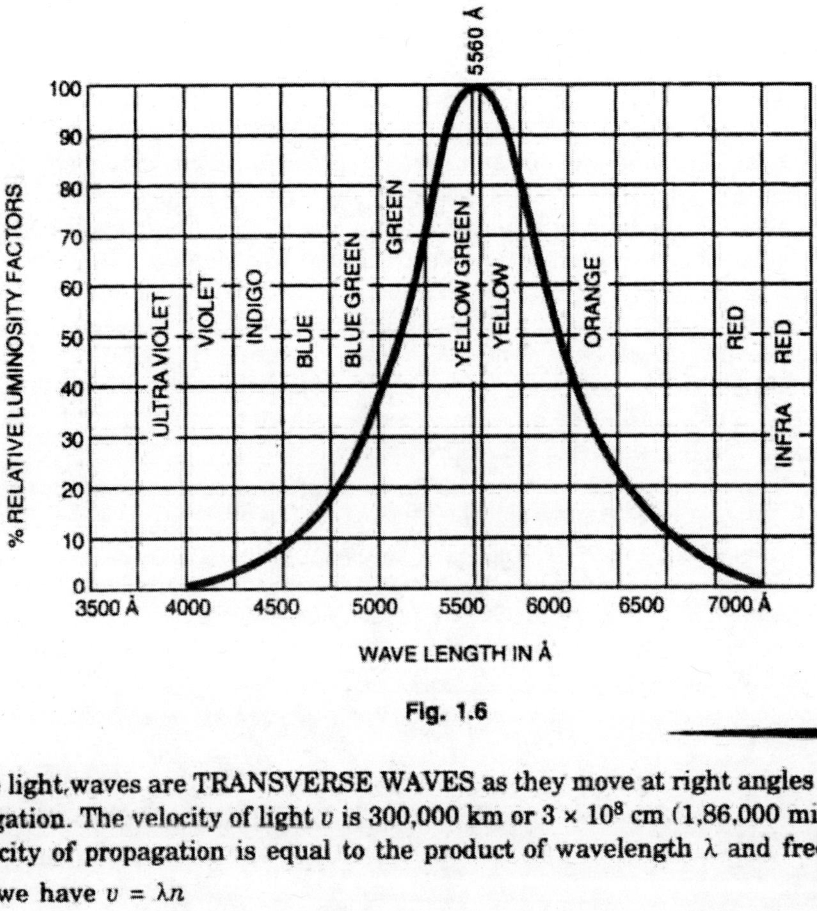

Fig. 1.6

The light waves are TRANSVERSE WAVES as they move at right angles to the direction of propagation. The velocity of light v is 300,000 km or 3×10^8 cm (1,86,000 miles) per second. The velocity of propagation is equal to the product of wavelength λ and frequency n.

So we have $v = \lambda n$

From above we know the values of v and λ and we have to find the value of n now.

$$n = \frac{3 \times 10^8}{5560 \times 10^{-6}} = 5.54 \times 10^{14} \text{ cycles per second.}$$

We have gone through the above exercise to understand more accurately the effect of the physical phenomena *i.e.*, the wavelength and frequency on the eye. In our day to day work almost unknowingly we come across so many wavelengths and frequencies and we restrict our vision only to the visible light that we can perceive. But before we produce the visible band of light between 3800 Å to 7500 Å we come across infra-red band and also beyond visible band, the ultraviolet band. Though these bands of infra-red and ultraviolet are not perceptable these cause great harm to our eyes. Care should be taken of eyes by all lamp makers to protect their eyes from exposures to infra-red and ultraviolet light.

The tube lights used in our homes actually produce ultraviolet rays between the end filaments. These ultraviolet rays are then obstructed by the fluorescent materials bonded on to the internal surface of the tube which break the ultraviolet rays to our visible spectrum band. Due to some material defects, sometimes after use even if a little portion of these coatings come off leaving the surface bare, there will be leakage of these harmful ultraviolet rays for which the tube should be taken out of service.

Colour Effects of Incandescent Lamps

In general, the eye performs well enough in the artificial light of an incandescent lamp which is nearer to the natural visible spectrum band. One of the advantage of the filament lamp is that the light it produces is quite uniformly spreaded in the visible spectrum. With the passing of the current, the filament becomes first red hot, then yellow and finally yellowish white. Actually the light of a filament lamp is a little yellowish white which being an easily available band no attempt is made to change this quality. Only for decorative purposes coloured shades, coloured glass shells and coloured coatings are used to serve the desired effect of light. The colour of the light is dependent on these extra application which practically do not require much change in the process of manufacturing of the lamp. While using internal coatings, the transparency of the shells and their capacity to withstand extra heat are of greatest importance. The glass shell makers should have a small batch of 'shells for coatings' tested before going into mass production.

However it is to be remembered that the filament of an incandescent lamp radiates in all directions at a constant level of illumination. The wattage of the lamps are also standardised and cannot be obtained as per the desired lighting power level. Almost at every hour there are high level illumination spots which cause great harm to the eyes. Also the brightness of the bare lamp strains the eyes very much. To avoid these difficulties of incandescent lamp a diffuser screen or concealed lighting system are being installed where continuous work in artificial light is to be carried out.

2

Artificial Light

The biological concept of living is based on the radiations received from the Sun—the life giver. A minute fraction of this radiated energy of the sun is perceived by us. Due to the presence of water, gases and minute particles in the atmosphere there are large variations and spectrum differences in the sunlight. However the first artificial light used for removing darkness at night was done by kindling wooden torches. The burning process in wooden planks, oil lamps or gases is rapid as the light is created by very fast oxidation of these burning materials. The heat here is concentrated on a very small area and the radiant energy starts flowing. Only a little part of this radiated energy becomes visible band and we term it as 'artificial light'.

The newtonian corpuscular theory was again revived by Max-Plank in 1900 which formed the basis of Einstein's photoelectric law. In incandescent lamps, the electrical energy is converted into visible light by quantum process when atoms of the conductor gets excited due to external pressure of the electrons which receive themselves of this electrical pressure by passing it on to the next group of electrons and the process continues. The energy thus produced is radiated externally and we call it as light when the frequency reaches the visible band. This type of energy radiation is known as 'quanta' or 'photon flow' which follow the known Max-Plank equation. While working with baffling problems of black body radiation Plank suggested a satisfactory solution of the law of radiation—That a vibrating electron emitting or absorbing radiation should possess energy as integral multiple of energy quantum.

$$e = h\upsilon$$

This is one of the most fundamental constant of nature. The Max-Plank constant is represented by the letter *"h"*. The value of *'h'* is very small. The energy produced by an atom is expressed in erg and it equals to 2.4×10^{-8} calories.

$$h = 6.6 \times 10^{-27} \text{ ergs per second}$$

The light follows Plankian constant $h\upsilon$ (υ is the frequency per second). For illumination purposes the electromagnetic wave theory of light is generally being followed for working with interference, diffraction and polarisation. The quantum theory of light is equally important and mentioned here as this might appear when making certain observations to study the characteristics of light relating to matter and radiation, Photoelectric effect, Raman effect etc.

The first artificial light through conversion of electrical energy was made by Humpery Davy in 1802 by producing an arc between two carbon rods. Light was produced but it had many disadvantages for putting it into general use. In 1841 de Moleyns took a British patent for making artificial light by putting a platinum metal foil which connected both ends of the poles but offered resistance to the flow of current. In America and Europe vigorous searches were made for a resistance material which would glow but would not burn out itself. On 21 October 1879, the great inventive wizard Thomas Alva Edison (*Photo 2.1*) of the United States made the first electric lamp using a few strands of sewing threads carbonised as filament. The first lamp burnt for 40 hours. This was the first practical incandescent lamp and Edison's name is immortalised with it. During the same period Joseph Swan of England made the same type of lamp but it's potentiality and seriousness were not envisaged by the British people and further development work was not carried out by them. This is the sad tale of a brilliant inventor who lost his legitimate prize due to lack of encouragement. Edison did not stop work with the invention of the '40 hour' lamp but put all his brilliance, strength and staked his wealth to prove the potentialities of this invention.

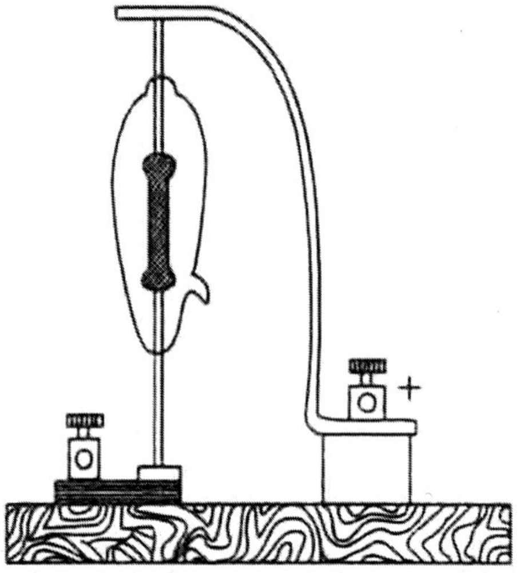

Fig. 2.1. The lamp made by Swan.

In 1878 Thomas Alva Edison joined the ranks of those striving for a practical incandescent lamp : Joseph Swan and St. George Lane-Fox in England ; William Sawyer, Albon Man, Hiram Maxim and Moses Farmer in the United States. The controversy over who in fact invented the incandescent lamp still rages today, especially the choice between Edison and Swan. Both made contributions—for instance, Swan insisted that carbon was the answer—but we do find it without question that Edison devised both a practical lamp of reasonable size and a complete electrical system, with both generation and distribution, to exploit his lamp (*Photo 2.2*). Because his first carbonized thread lamp ran 40 hours before intentional overvolting destroyed it, and because this happened on 21st October 1879, we take this as the starting point for a century of active lamp research and development, a century rich in interplay between physics and technology.

Edison had realized that parallel wiring, where the current and not the voltage is subdivided, was the only practical system in which lamps or other current-consuming devices could be separately switched on and off and which could thus lead to successful distribution. Edison's lamps were adapted to this system, having long thin filament wires to give them a high electrical resistance. This essential idea distinguished his lamps from those of Swan and others, who were still working on the ideas of series wiring, as was set down by Edison in his most important patent, lodged on 10th November 1879. Under Edison's scheme the current from an electricity plant, where several dynamos, according to requirements, could be connected in parallel, would be carried by insulated copper rods, tubes or cables laid in the streets, and thence conducted by thinner branch lines into factories, shops and houses, just as in the case of the gas mains from the gasworks. Edison had to design every component of the installations required for this scheme. The lamp cap (or base) by which millions of electric lamps are still today fixed in their holders is the 'Edison screw cap'. He invented the 'fuse', which automatically breaks the circuit in any branch of the network where

T.A. EDISON.

Electric-Lamp

No. 223,898 Patented Jan. 27, 1880

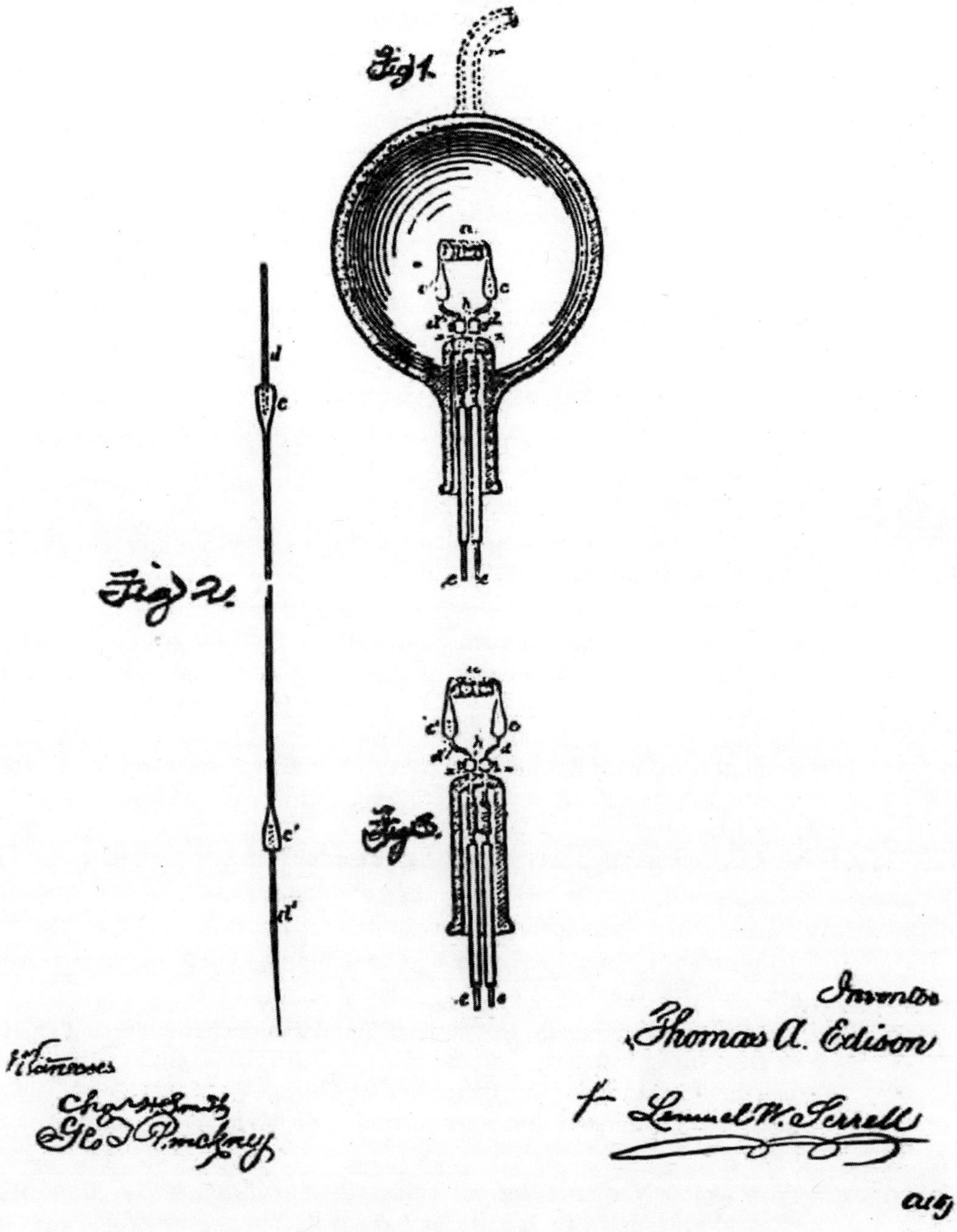

Fig. 2.2. Drawing on application for patent of electric lamp by Edison. (Courtesy : G.E.C. Cleveland, Ohio)

serious overloads or short-circuiting occur. He improved the magnetic circuit of dynamos to such an extent that their efficiency rose from about 50% to 90% and more. There was also an electric meter, which measured consumption by weighing electrolytic copper deposits, among the series of inventions with the aid of which Edison at the end of 1879 was able to present the world with a serviceable system of electric lighting.

While Edison was pushing ahead in the United States with the manufactures of all those components for his installations, he attracted attention to his system in Europe by a spectacular exhibit at the first International Electric Exhibition at Paris in 1881 (where incidentally several other electric lamp manufactures including Swan, Lane Fox, Maxim and Siemens had their products on display) (*Photo 2.3*). Edison demonstrated there a lighting system comprising 1000 lamps of 16 candle-power each, fed by his first large dynamo which, with its associated steam engine, weighed 25 tons and delivered 70 kW. His lamps at that time were already in mass production—in the first 15 months he had sold 80,000 at an initial price of about 3 dollars each. His success is now shared by the whole mankind; a simple touch on a switch and the darkness is removed !

Chronological Development of Early Lamp Making

1879. The first incandescent lamp was made by Edison with carbonised filament, a few strands of sewing threads were carbonized in an air-tight crucible. Light strength was 1.4 lumen per watt.

1881. The first commercial lamp made available with 1.6 lumen per watt used carbonised bamboo as filament.

The world of electric lamp is also much appreciative of the young visionery and energetic engineer Gerard Philips (*Photo 2.4*) who set up his factory Philips and Co. at Eindhoven, Holland in 1881. The very first consignment of electric lamps made by Philips was sold to a candle factory a competitor in the field of lighting who purchased the lamps to overcome fire hazards. After a century of devoted work with lighting, the name of Philips is synonym with electric lamp (*Photo 2.5*) though they produce various other electrical and electronic quality products.

1883. To increase the life of the filament—absorbent cotton dissolved in zinc chloride molasses was passed through a die for uniform thickness and shape which was then put into alcohol bath. The cellulose string thus made had good strength. After mounting this filament inside the lamp, the lamp was aged in hydrocarbon vapour. This filament gave 3.3 lumen per watt.

1893. The first metal filament, the 'stopper lamp' was made by Westinghouse company of USA using soft iron. Two hundred and fifty thousand of these lamps were switched on at Columbian Exposition at Chicago on May 1, 1893. The lumen per watt was 3.8. For the first time alternating current was used to light an electric lamp.

1897. Prof. Walter Nernst of Germany made the metallised carbon filament named as "Nernst Glower" with Ziconia as base with Yttria and Erbia. This lamp gave about 4 lumens per watt.

1905. In Europe, the osmium filament giving 5 lumens per watt was introduced this year. Osmium is a very expensive rare metal and the filament was found to be very fragile.

1905. Tantalum filament was introduced in the market and continued to be a favourate for five to six years though the metal got fractured due to crystallisation when alternating current passed through it.

1907. The first use of tungsten filament was started by Franz Hanamam and Alexander Just. The tungsten metal was powdered, made into a paste with a binder producing tungsten thread. After removing the binder the long thread was supported by hairpin loops. The lamp gave 7 lumen per watt.

1911. High standard of purification of tungsten was achieved. Coolidge discovered the process of making tungsten very pure and ductile by working it below annealing temperature. Tungsten was drawn into a very fine wire, better light and longer life was obtained. The lamp gave 10 lumens per watt.

1913 to 1917. Till now the lamps were evacuated of air and known as vacuum lamps. To get better performance from the filament Langmuir used still thinner filament and double coiled it. This concentrated the heat, the surface for emission was restricted to a very small area. The filament being very thin and the heat so high, the material got evaporated very quickly. To overcome this difficulty Langmuir first evacuated the lamp and then introduced an inert gas which became the forebearer of the present day lamp. This filament gave 14 lumens per watt which was really a great achievement.

From 1913 till today innumerable varieties of lamps have been made to provide light at particular demands but the basic principle of incandescent lamp making have remained the same *i.e.,* to put resistance in the flow of electricity, the conductors of which are enclosed in an evacuated or inert gas filled atmosphere.

1934. The single coil filament has been replaced by coiled-coil filament by Benbow to improve the efficiency of the lamp by 15%.

Tungsten is still the main metal for filament in incandescent lamp. The melting point of tungsten is 3380°C and at this temperature it gives 52 lumens per watt. Continuous researches are being made to reach as high temperature as possible to get maximum lumen per watt and at the same time to reduce the evaporation quality of tungsten for prolonging the life of the lamp. Presently the highest achievement have been seen on a 'photo flood' lamp giving 35 lumens per watt with 3 hours burning life. The following details show how vast is the field of lamp making and the widest range of it's application.

The tiniest lamp made today is known as 'gram of wheat' used for intestine inspection in surgical instruments. This is only 0.34 inch long and 0.08 inch in diameter producing .35 lumen per watt and operated by a dry cell consuming 0.17 watt current. It weighs only 1/756 lbs.

The largest lamp is 35 inch high, 20 inch in diameter and weighs 35 lbs. It is a 50,000 watt lamp producing light of one thousand 100 watt lamps. The filament weighs 1.6 lbs. The filament being so large and the working wattage being so high, the filament gives 32 lumens per watt.

So, at one end there is the tiny 0.17 watt lamp and at the other end it is a 50,000 watt lamp—how large the range is ?

1958. Mention also must be made of the halogen lamp first made by Elmer Fridrich and Emmett Wiley which is an improved version of the incandescent lamp. The process of evaporation of the tungsten filament could not be arrested fully at high temperature, but the evaporated tungsten could be reinsteaded on to the surface of the filament by introducing a very minute quality of the elements from halogen group. The halogen lamps may prove to be the lamp of the future.

1982. In Great Britain, M/s. Thorn Lighting Ltd. has claimed to have made a great improvement on incandescent lamp. A lamp producing 1400 lumens (equivalent to 100 watt) would consume only 20 watt of electricity and it's burning life is extended to 5000 hours form the normal life of 1000 hours.

BROAD FIELDS OF LAMP USAGE

Multi-various types of lamps are (including gas-discharge types) manufactured mainly to suffice the following purposes.

1. **For Producing Heat.** The filament produces first heat and with further pressure from energy, it produce light. In the fields where dry and localised heat is only necessary, lamps are used in which the energy is fully utilised for producing heat. These lamps have many advantages over conventional bare nichrome wire heaters and are in great demands now in medical treatment, painting industry and leather industry. The metallic filament in these lamps radiate mostly 'Infrared Ray' and are called "Infra-red Ray Lamps".

2. **For General Lighting Purposes.** Removal of darkness or as indicators.

This is the main group in which almost ninety per cent of the lamps are made :

(*a*) G.L.S. (General Lighting Service) lamps are made for service in homes, streets and factories etc.

(*b*) Auto Lamps—for automobiles.

(*c*) Mini Lamps—for decoration, indicator etc.

(*d*) Projection, Photoflood, Photo Flash Lamps.

(*e*) Search Light, Landing Lamps.

(*f*) Halogen Lamps.

(*g*) Mercury Vapour Lamps.

(*h*) Aviation Lamps.

(*i*) Sodium Vapour Lamps.

(*j*) Xenon, Krypton Lamps.

(*k*) Instruments and Machine Lamps.

(*l*) Neon Signs and Tubes.

(*m*) Fluorescent Lamps.

(*n*) Surgical, Medical Lamps.

(*o*) X-Ray Tubes etc.

Photo 2.1. The wizard of many inventions. Thomas Alva Edison in around 1883 with some of his special bulbs.
(Courtesy : GEC. Cleveland, Ohio)

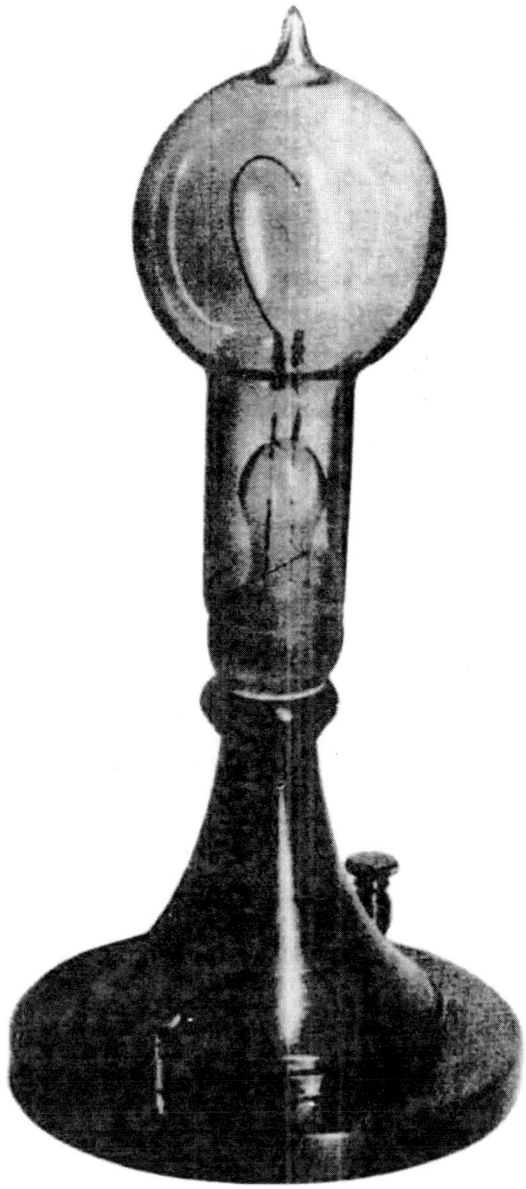

Photo 2.2. Edison's lamp.
(Courtesy : General Electric Co., Cleveland, Ohio)

Photo 2.3. Collection of 20 incandescent electric lamps, bought by the Teyler Foundation in 1881 at the Paris exhibition, including lamps made by Edison, Swan, Lane-Fox, Maxim and Siemens.
(Reproduced by Courtesy of Teyler's Museum, Harrlem and Philips Tech. Review).

Photo 2.4. Gerard Philips a stalwart in electric lamp technology experimenting in his home-made laboratory at Zaltbommel in 1890.
(Courtesy : Philips Tech. Review, Holland).

Photo 2.5. A modern lamp factory hall with a number of production groups parallel to each other in the Philips Lamp Works at Weert (Netherlands). Each production group covers the whole width of the hall and constitutes an independent unit. Each group, which is operated by three or four persons, is supplied with tungsten filament, support rods, bulbs, lamp caps etc., and delivers a continuous stream of finished incandescent lamps (more than 2000 per hour). Here too the lamps of each production unit are subjected to the necessary inspection. The stream of lamps coming from each unit is carried on a conveyor belt through the floor to the packing machines under the hall. (Courtesy : Philips Tech. Review, Holland).

3. **For Ultra Violet Working Purposes.** These are all special purpose lamps and a very good scope for manufacturing exists in India in this field. The ultra violet ray is produced by exciting the gases inside the lamp through high volt electrical discharge. The following lamps come under this category :

(*a*) The Sun Tan Lamp

(*b*) Germicidial Lamp

(*c*) Black Colour Lamp

(*d*) Photochemical Lamp.

4. **For Energy Conversion Work.** These are also special purpose lamps where light energy is being converted into mechanical or chemical purposes and are known by their use :

(*a*) Photoelectric cells—safety, burglar alarms etc.

(*b*) Lamps for plant growth

(*c*) Lamps for attracting phototropic insects

(*d*) Lamps for fading of colours, dying and bleaching.

5. **For Electric Purposes.** Certain crystals and liquids emit fluorescence and this phenomenon is utilised in the electronic industry for the visible displays. A very vast scope exists in India in this field.

A very efficient way of generating hole-electron pairs is by carrier injection across rectifying junctions. Oleg Loessev discovered light production by this phenomenon in silicon carbide as early as 1923. Efficient light generation here requires high injection ratios, a high proportion of radiative recombinations of hole-electron pairs, and rectifying structures that enable escape of light. Electroluminescent light-emitting diodes embody such structures. These diodes pervade light in the displays of electronic wrist-watches and pocket calculators.

Though not cheap or efficient enough for general illumination, light-emitting diodes have adequate quantum efficiencies for use in indicator lights and in injection lasers in gallium arsenide and similar materials. (Quantum efficiency is the ratio of the number of photons emitted by a material to the number of electrons or ultraviolet photons incident on that material). Another type of indicator lamp combines phosphors that convert infrared radiation into visible light with gallium-arsenide light-emitting diodes. These devices convert 0.9 micron radiation in the infrared region into visible light. The two light sources used in the electronic industry are :

(*a*) Light Emitting Diodes (LED).

(*b*) Liquid Crystal Displays (LCD).

EFFICACY OF LIGHT SOURCES

Edison's early incandescent lamp was far from ideal. This lamp gave 1.4 lumen per watt whereas a present lamp giving 17.4 lumen per watt also is considered "lossy" because about 80 per cent of the lamp's energy is radiated in non-visible wavelengths. Presently 'lumens per watt', defined as the lamp's "efficacy" (formerly termed as efficiency) is taken as a measure of the lamps efficiency as a light source.

The following graph gives approximate efficacies of various light sources.

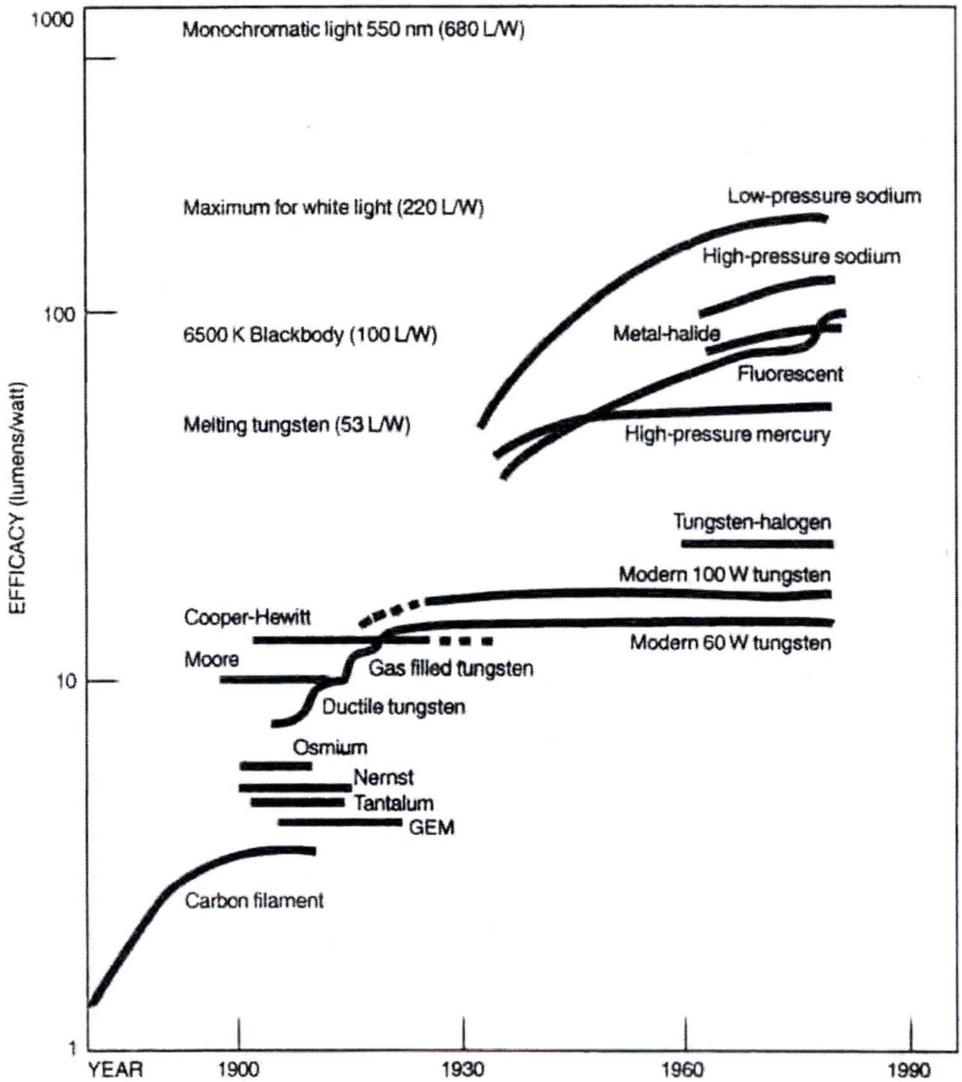

Fig. 2.3. Approximate efficacies of various light sources. The efficacies are plotted as a function of time for different kinds of electric lamps showing the evolution within each technology as well as progress from one technology to another. This also indicates efficacies of several standard sources.

(Courtesy : GEC, Cleveland, Ohio).

3

ELECTRIC LAMP TECHNOLOGY

An 'electric lamp' is a contrivance which converts electric energy into 'Light energy'. With the help of electricity, artificial light is produced mainly in the following process :

Artificial Light
(Produced by Electricity)

DIRECT
Filament Incandescence
GLS, Auto, Mini,
Halogen

DIRECT
Electroluminescence
Semi-Conductor

INDIRECT
Gas Luminescence
Fluorescent, Metal
Hallide, Sodium,
Mercury Vapour

INDIRECT
Electroluminescence
Air Craft, Panel Lamp
LED, LCD

A 'tungsten filament lamp' produces electric light through 'Direct Filament Incande Scence' hence it is also called an incandescent lamp.

When a bare resistance wire is placed in the path of electricity, it get's heated up-starts glowing giving light and after some time it burns out. The reasons for its burning out are (*a*) the weak properties of the resistance wire to withstand heat (*b*) the presence of oxygen surrounding the resistance wire, which oxidises it very quickly.

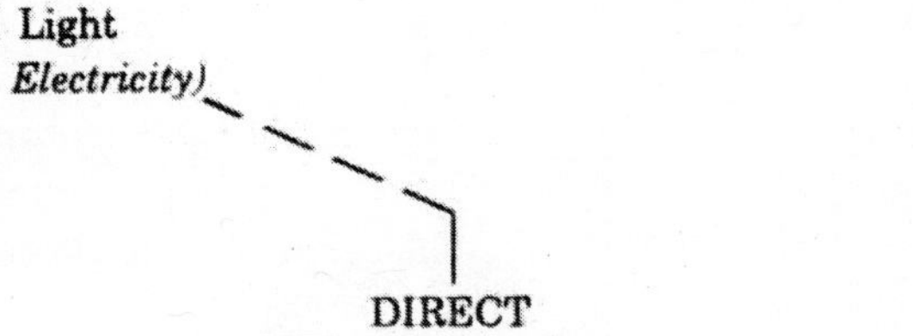

Fig 3.1

So, to produce stable light, the very first problem was to find the proper material for the resistance. The material should have all the good qualities of a very tough metal—with highest possible melting point, lowest co-efficient of expansion and tough bonding characteristics even after burning. Tungsten was found to have all these qualities and till now it is the best metal used as resistance known as filament. (See Chapter 4.)

The next problem was of burning out of the filament due to the presence of oxygen in the air. So, the resistance is enclosed inside a glass envelope and the air is evacuated.

After evacuating the glass bowl, it is seen that at high temperature, the tungsten starts evaporating quickly. The vacuum around the filament augments evaporation. The behaviour of molecules under vacuum or partial vacuum conditions are better experienced on hill tops and mountains where the atmospheric pressure is less—meaning a partial vacuum condition. Here water boils below 100°C because the hot molecules of water vapour on its surface are being sucked away by the atmosphere.

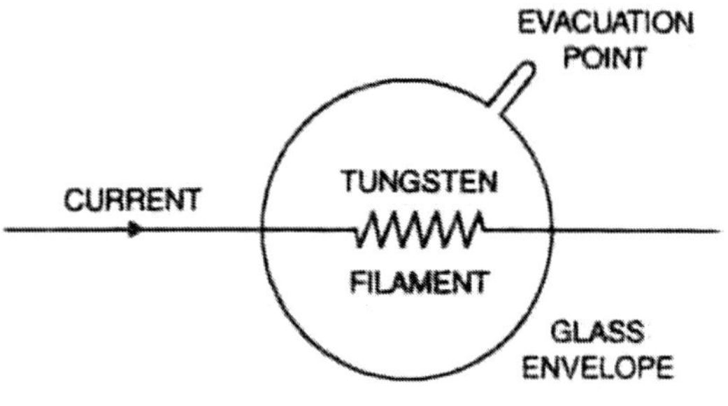

Fig 3.2

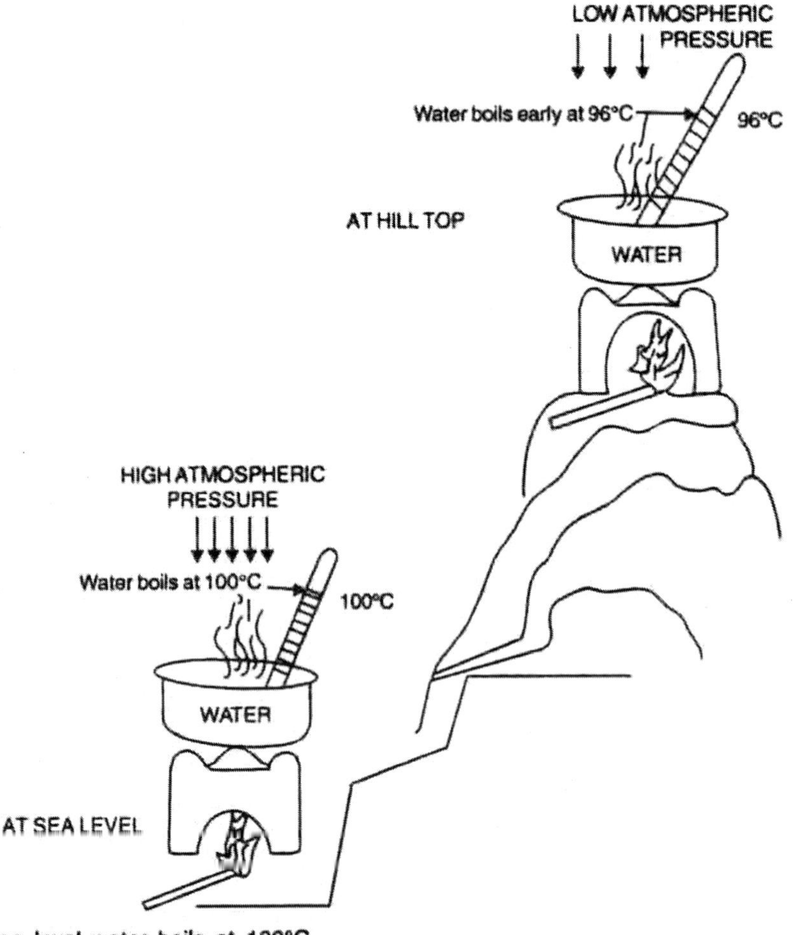

On sea level water boils at 100°C.

On hill top water boils below 100°C (at 96°C).

Fig 3.3. Effect of external pressure (here atmospheric pressure) on dismemberment of molecules.

Similarly, under vacuum condition the tungsten inside the glass bowl of a lamp evaporates very quickly.

The evaporated tungsten gets lodged on to the wall of the glass bowl blackening its inner surface and finally the filament breaks.

Problem 1. To produce more light more heat is needed. When more heat is created inside the vacuum, the tungsten gets rapidly evaporated.

2. Inside the lamp though the resistance burnt and produced light, less light could pass through the blackened wall of the glass bowl. So, more light could not be obtained even after utilising more resistance.

Solve 1. The rapid evacuation had to be reduced by creating some means inside the glass bowl to decelarate the evaporation process of tungsten.

2. The filament had to be so designed that it would create and retain maximum heat around it to produce more light with reduced blackening of the wall.

It is known that the boiling point of water in a boiler is more than 100°C. The reason being that the steam inside the boiler creates a heavy pressure on to the surface of the water and does not permit it to "steam off" easily (Fig. 3.4).

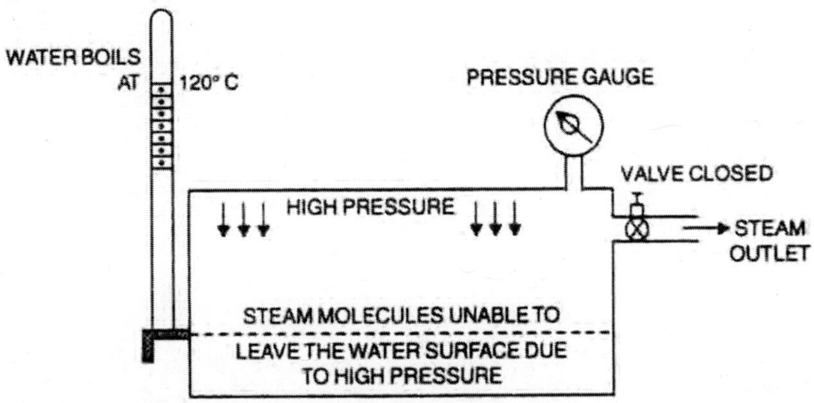

Fig. 3.4

The water lets off steam of 100°C as usual but the newly evaporated water molecules cannot leave the surface due to the earlier produced steam which is not permitted to escape due to the closed valve but being utilised to create positive pressure on the water surface so that the newly evaporated molecules fall back into the water again.

The same principle is used in lamp technology. argon, nitrogen, krypton, xenon—the inert gases were found suitable to be introduced inside the lamp as these gases do not react with any of the components inside the lamp.

The inert gases produced positive pressure over the filament. As the filament glows and becomes warmer and warmer its burning surface gets loosened and start throwing away the loose tungsten metal. The pressure being high over the surface made by the inert gas, the loose metal cannot 'fly off' easily.

Introduction of inert gas undoubtedly increases the filament life but brings in numerous problems.

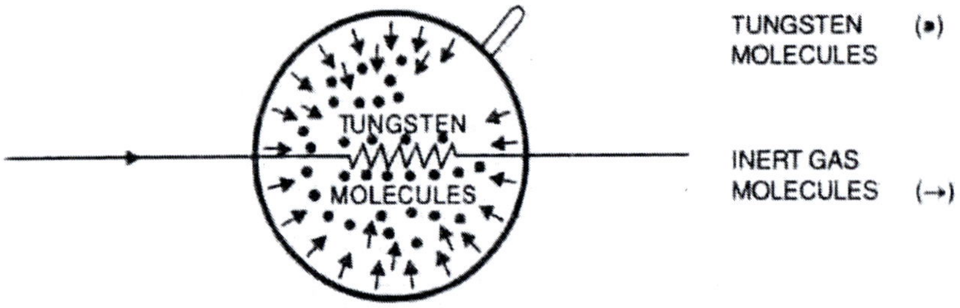

Fig 3.5. Inert gases producing pressure over filament surface.

1. In a vacuum lamp, the heat from the filament could dissipitate by radiation and conduction through electrodes. So retention of heat was more on the unit surface of the filament. The long filament had to be spreaded in many position around the centre of the glass shell. The size of the glass shell also did play a great part.

In vacuum there was no medium to carry the heat away. After introduction of gas, the foremost problem is of heavy heat loss by conduction through them. Naturally the light output was very poor though the filament life increased.

Solve. Keeping the same resistance value, the filament is twisted and the same length of wire now occupies less space and the heat is concluded within a limited area. By further twisting of the wire, small coils are produced which do not touch each other. Through proper arrangement this shape of the coil has been arrested and named as single coil filament.

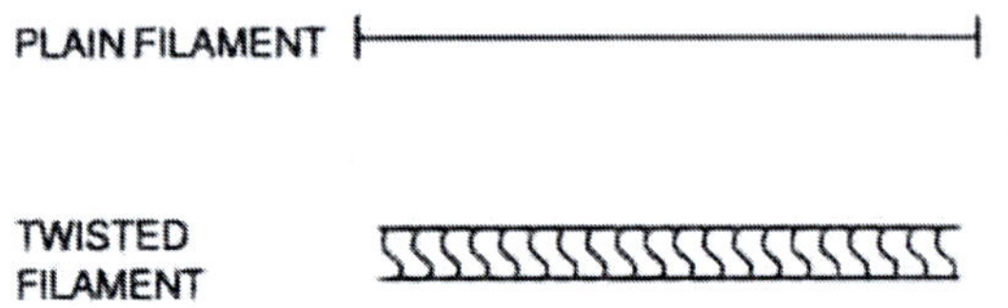

Fig. 3.6

The efficiency of light was further improved by coiling this single coil of the twisted filament. Being shorter in length the complete heat from the filament was available near the centre of the lamp shell within a short space inside the lamp.

So far lamp technology developed and proceeded mainly on trial and error basis. After introduction of single coiled filament great sophistication and scientific calculations became more pronounced in the manufacture of electric lamps. The theory of the 'hot gas sheath', the introduction of inert gas and shortening of the plain filament by coiling (partly) are the contributions of the mobel laurate Irving Langmuir which are the basic principles in lamp technology.

Invention of Irving Langmuir. The first electric lamp was invented by Edison in 1879. The invention of "stationary sheath" principle by Irving Langmuir (*Photo 3.1*) in the year 1912 is the next most prominent landmark in the technology of lamp making. Langmuir enumerated that the viscosity of a gas increases with increasing temperature—and a stage reaches when this viscosity becomes so high that this gas becomes stationary around the incandescent body. Normally the hot gas always gives way to the cooler gas but at high temperature, this flow of convection totally stops and the heat transfer occurs only through conduction. For easier understanding, it is better to assume that every incandescent body is always surrounded by a layer of gas. Consequently there will be temperature gradient of the viscous gas adjacent to the incandescent body and the gas lying away from it.

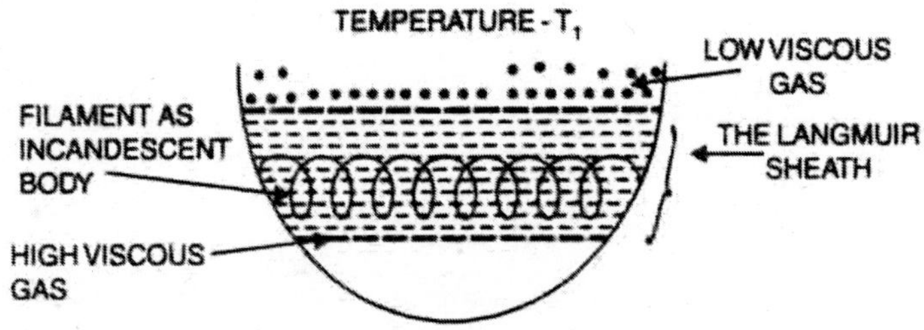

Fig. 3.7

In a lamp where the tungsten molecules from filament evaporate they travel upto the edge of the sheath with a good thrust but loose it's energy when they meet cold and denser gas. Without finding any escape route they again return to the filament reducing the evaporation process. The rate of evaporation has been established by G.R. Fonda as following :

$$maP \log b/a$$

where m = rate of evaporation

P = pressure

a = diameter of the filament

b = diameter of Langmuir sheath.

THE LANGMUIR SHEATH

Langmuir stated that :

1. The loss of heat from wires by free convection takes place exactly as if there were a film of stationary gas around the wire, through which the heat is carried entirely by conduction.
2. The thickness of the film is independent of the temperature of the wire, but probably increases with increasing temperature of the surrounding gas.
3. The thickness of the film of gas varies in a single way with the diameter of the wire, namely.

$$b \ln \frac{b}{a} = 2B$$

where B being a constant for any gas

b is the diameter of the film of gas

a diameter of wire

ln is the length of the incandescent wire.

This is known as the Langmuir's formula as used in the electric lamp industry.

4. The rate of convection of heat from any wire is equal to the product of two factors :

(*a*) The shape factor expressed as S which involves only the diameter of the wire

with

The constant 'B' for any gas

(*b*) The heat conductivity of gas as a function :

Thus W is the energy loss from wire in watts per cm ψ (Psi)

$$W = S\,(\psi_2 - \psi_1)$$

where S may be found from the equation

$$\frac{S}{\pi} \cdot E^{-\frac{2\pi}{S}} = \frac{a}{b}$$

and
$$\psi = .419 \int^{T} KDT$$

K is the heat conductivity of the gas at the temperature T in cal/cm°C, ψ_2 is taken at the temperature (T_2) of the wire, and ψ_1 is taken at the temperature (T_1) of the atmosphere.

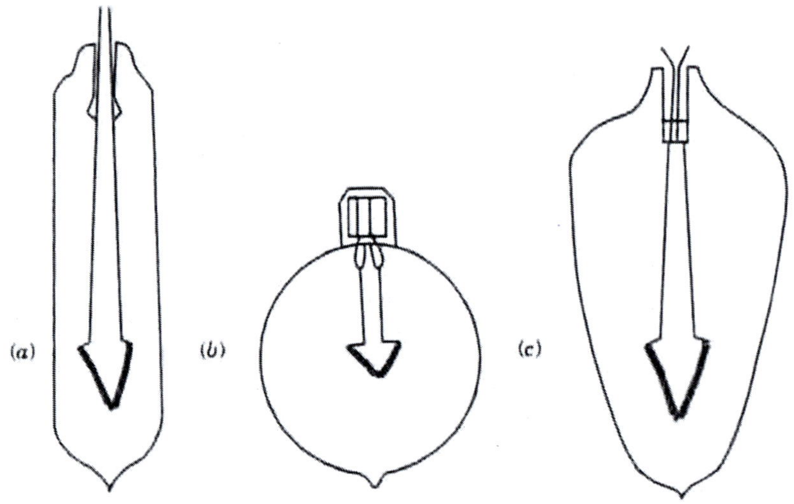

Nitrogen filled lamps with coiled filament patented by Langmuir in 1918.

Fig. 3.8

Comments :

1. All the lamps have single coil filament.
2. In the special shape lamp (*b*) the filament is placed at the geometric centre as in modern lamps.
3. The lamps (*a*) and (*c*) are lower voltage lamp in which the filaments are kept at a greater distance from the outer connection to avoid loss of heat through connection.

By an ingeneous method the coiled filament is again spiralled so that the length of the filament is reduced to its minimum and the heat concentration is maximum at the centre of the shell.

Comparison between Single Coil and Coiled Coil Filament

Introduction of inert gas caused great loss of heat and to overcome it, coiled coil filament was introduced. The loss of heat inside a lamp may be calculated by following the formula enunciated below :

$$W = C \times l \times d\,(\delta) \times T(\tau)$$

where W = loss of heat to the gas

l = length of the filament

(δ) d = diameter of the filament

(τ) T = absolute temperature of the filament

C = constant of the gas mixture.

The values of exponents δ (delta) and τ (Tau) have been determined through many experiments as following

$$\delta = 0.3 \qquad \text{and} \qquad \tau = 1.6$$

By elaborating the above formula, it has been proved theoretically that the length of the filament could be reduced to half of it's length by coiling it around a mandrel giving a coiled coil shape. On an average a 15% efficiency was obtained over single coil filament as shown in the following graph.

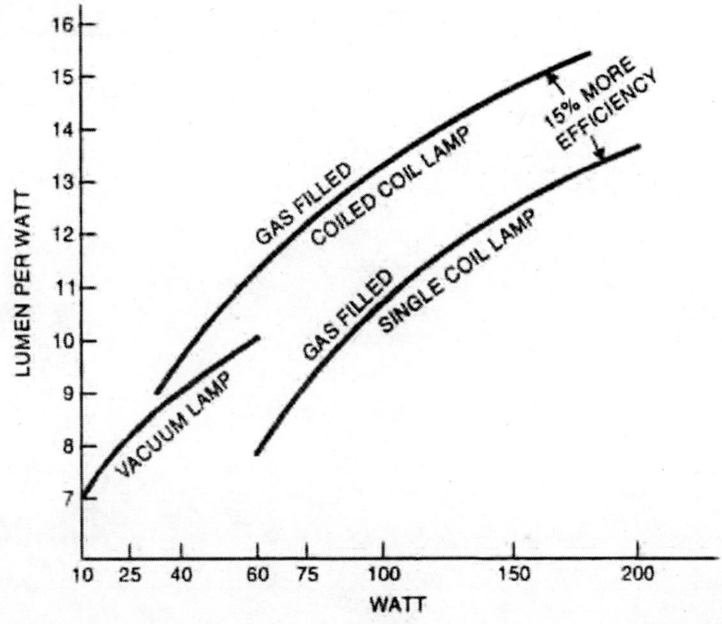

Fig. 3.9

For concentration of heat, the size of the shell and the placement of the filament at the centre becomes then very critical and of great importance.

By coiling a wire, the area of free surface responsible for mass losses is reduced to about half which simultaneously also reduces the loss to half.

The weight loss of a coiled coil filament in gas filled atmosphere during the total life has been established to be 2% of the original filament.

The research and development of incandescent lamps are mainly concentrated on the filament and the inner atmosphere as almost all the other attributes have reached to their optimum efficiency.

The Lamp Centre Length

This is the distance from the geometrical centre of the filament to the contact plate of the cap including the height of the solder. The easiest way to measure light centre (LC). Total length of the Lamp (L) minus diameter of the shell (d)

$$LC = L - d.$$

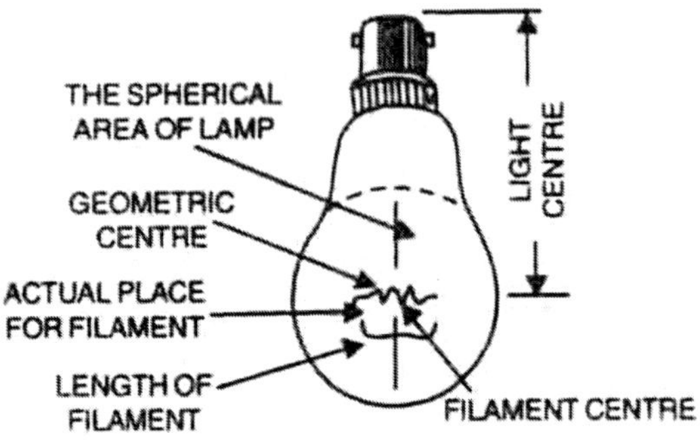

Fig. 3.10

The centre of the filament should lie at the geometrical centre of the shell. In lamp making this is generally not given much importance. The effect of off-centred filaments are shown below :

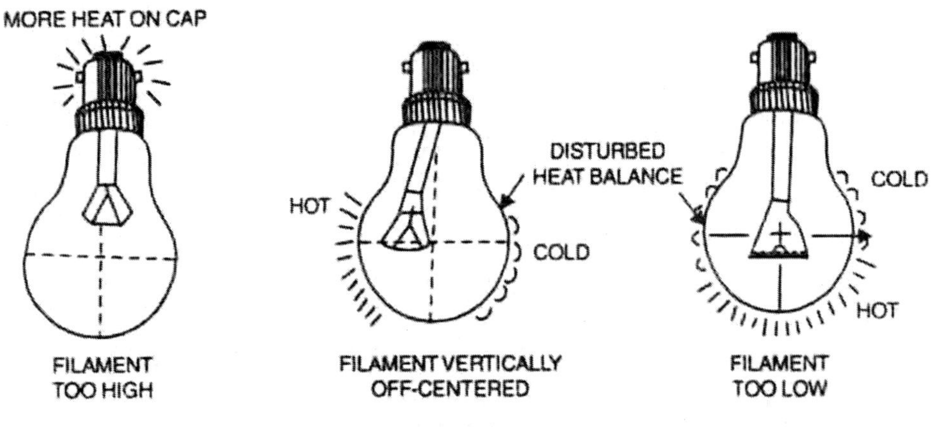

Fig. 3.11

In automobile headlamp the lamp centre plays a great role. For gathering maximum power from the auto-lamp and at the same time to avoid spherical aberration of light, a parabolic mirror is used.

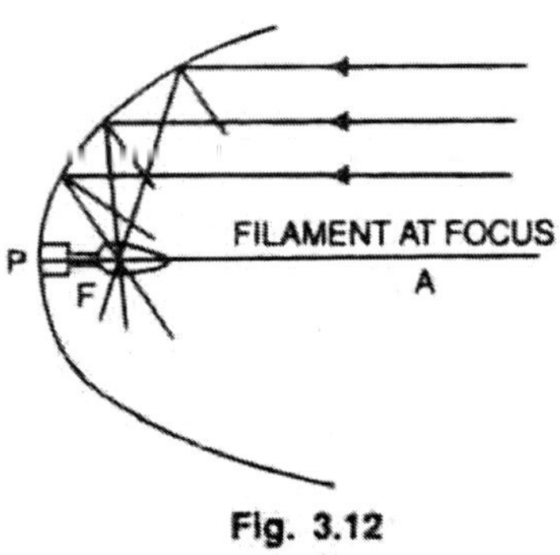

All the rays emanating from the autolamp filament at the focus *F* of the parabolic mirror (*P*) are reflected parallel to the axis (*PA*)

Fig. 3.12

If the filament '*F*' is even slightly disturbed—nearer or away from *P*, the whole geometry of light will fail—the light from the automobile will either blind the driver of the approaching vehicle or fall flat on the road just below the head lamp serving no purpose. So the lamp centre length is statutory measurement for the makers of autolamps.

The Shell. The optimum or the best possible glass shell sizes have been found out from the tested formulas, and their practical applications. The following gives an example how improvement on glass shell diameter and centre length have been reduced to the present size to concentrate heat at the filament centre without sacrificing the span of the burning life.

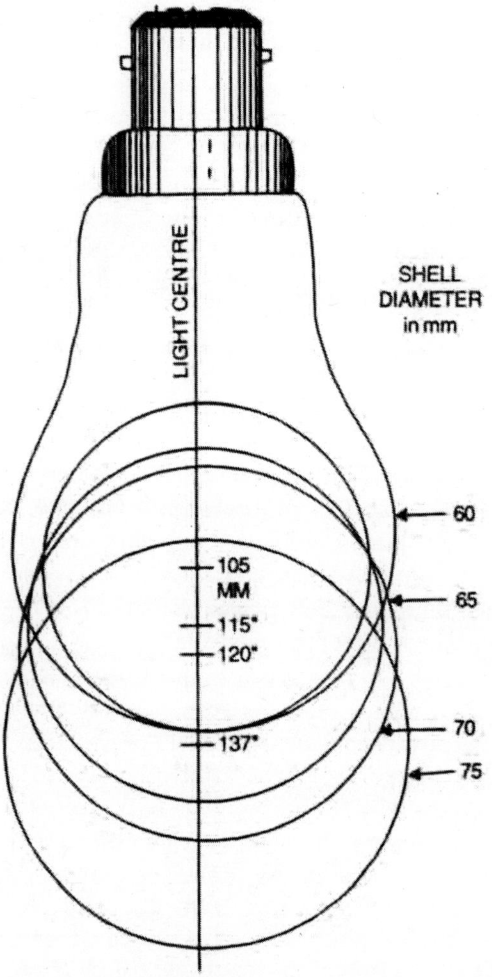

Fig. 3.13

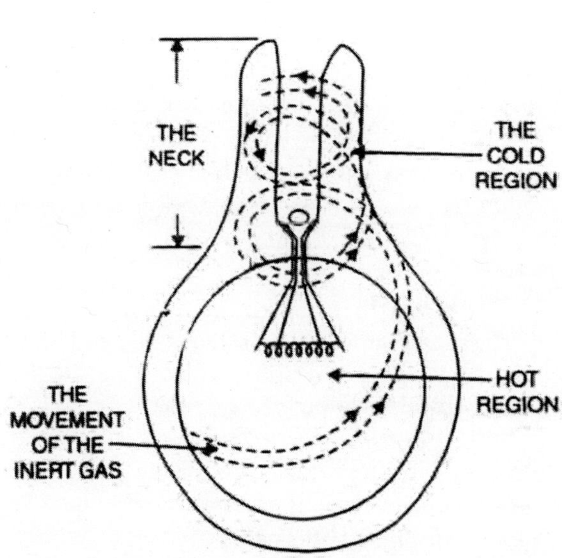

Fig. 3.14. The neck.

Year	*Glass Shell dia (mm)*	*Centre Length (mm)*
1943	75	137.5
1955	70	120
1956	68	115
1961	60	105

Many trials have been conducted to reduce further the glass shell diameter of *GLS* Lamps than the present size (less than 60 m/m as in the above case of 100 watt lamp). Main problem is found to be excessive rise in lamp cap temperature which is now brought under international standards.

The Neck. This constitutes also a very important part in the lamp. The elongated area away from the spherical shape act as a recess for the evaporated tungsten molecules from the filament. The spherical portion of the glass shell is comparatively hotter being nearer to the filament and the area in the neck is comparatively less hot Fig. 3.14. So, there is a convection current forming it's colder vortex inside the neck. The neck also keeps the lamp cap having connection with the mains supply at the optimum distance to avoid overheating. The neck cannot be made excessively long as it will reduce the filament temperature.

Inert Gas. By using a mixture of nitrogen—krypton, this rise in temperature can be overcome, but the excessive cost of krypton, gas does not make it economically viable. Abundantly available, the cheapest inert gas is nitrogen. It's molecular weight being low, it cannot put much pressure on the disintegrating molecules of the tungsten filament specially at higher voltages. The next choice fell on argon which gave three times more life than the nitrogen filled lamps. Use of pure argon in lamps again brought some difficulty at higher temperature. Argon molecules in combination with very minute particles of impurities, produce electrical discharges between exposed portion on the filament at the foot of the stem. So, a little amount of nitrogen gas is used to avoid this local action. For higher wattage lamps, 92% argon with 8% nitrogen gas mixture is recommended. For lamps upto 100 watt, a gas mixture of 86% argon and 14% nitrogen is more economic and efficient too.

From experiments it has been found that vacuum lamps is Mini group could be made of upto 6 watt power—for higher wattage lamps, these should be gas filled.

G.L.S. lamps, as night lamps and decoration lamps of low power consumption upto 15 watt can be of vacuum construction. Beyond this they should be gas filled. In India, upto 25 watt lamps are of vacuum type but these will soon be also brought into gas filled category.

Holographic Interferometry. In 1973 E. Fischer and J. Fitzerald of Philips (W. Germany) made a detailed study of the tungsten evaporation (transport) process within a gas filled lamp. *(Philips Tech. Rev. 35.297, 1975). They used tracer techniques in Holographic Interferometry.

The main criteria of the experiment was to determine the temperature and the flow field of gas filling which ultimately govern the transport process. As there are innumerable shapes and designs of lamps, they selected a cylindrical lamp with 4 cm dia. and 15 cm length and made different experiment by altering the diameter. The wire temperature was measured by pyrometric connections through a slit in the tube. The result obtained through pyrometer and the photographic plates were fed to a computer for numerical solution of equations in fluid dynamics. From this the following temperature distribution (left) was obtained and with the tracer particles of quartz glass (25/mm) the interference pattern (right) was obtained.

[*See Photo 3.2, 3.3 (a) and 3.3 (b)*]

This not only proved that the calculated and actual behaviour of the isotherms were similar in pattern and their behaviour but the changes also were found concurrent under different pressures of argon. As the pressure is increased, the layers of isotherm increase, flatter out and put more pressure at the point of thermal origin. The pattern suggests that the vaporisation of tungsten will be reduced due to excess pressure of argon over the hottest zone inside the lamp.

SOME INHERENT PROBLEMS OF LAMP MAKING

More heat from the filament produces more light with the sacrifice of it's longivity. Incessant efforts are being made to bridge the gap between the light and life of the filament. Presently the researchers are confronted with the following inherent problems :

1. **Thermodiffusion of the Filling Gas.** The hottest portion in the lamp is the filament and the atmosphere just adjacent to it. The shell body which comes into contact with the outside atmosphere and being farthest from the filament is the coldest portion. It creates a considerable temperature gradient in the gas inside the lamp. Thermo-diffusion is that characteristic of the gas which causes the heavier molecules to diffuse towards the cold region and the lighter molecules covering the hottest region. This means that the average density of the gas atmosphere around filament is less than the original pressure used while filling. The filling of only argon gas is restricted due to it's sparking prunness at high voltage. Excessive use of only nitrogen again induces inefficiency through thermodiffusion. For this reason, the maximum permissible percentage of heavier gas argon is 92% and nitrogen 8% in argon-nitrogen gas mixture which gives the best result.

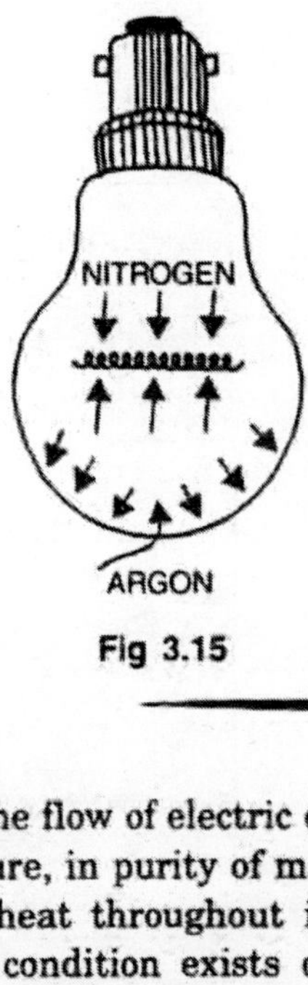

Fig 3.15

2. **Hot Spot.** A filament is a simple resistance in the path of the flow of electric current which produces heat. Even after taking greatest care, in manufacture, in purity of material, drawing it to size and shape, a filament cannot produce equal heat throughout it's full length. So, there is temperature gradient wherever sub-normal condition exists causing constriction to the flow of current. At these places of constrictions, the resistances become higher producing more heat which cause more transportation of tungsten molecules making the filament thinner and thinner on usage. These are known as 'hot spots' from where the filaments finally give way.

Later on 'potassium bubbles' have been inducted into the tungsten molecules for interlocking of the crystals to avoid sagging of the filaments. Some 'hot spots' may be attributed to the potassium bubbles. Also variations in the filament shape and roughness of the surface cause hot spots which can be overcome by proper manufacturing and working process.

The surface roughness causes variations in the emmissivity and the wire diameter. Even an initial temperature difference of 5 K along the local areas ultimately produce 'hot spot' which determine the life of the lamp.

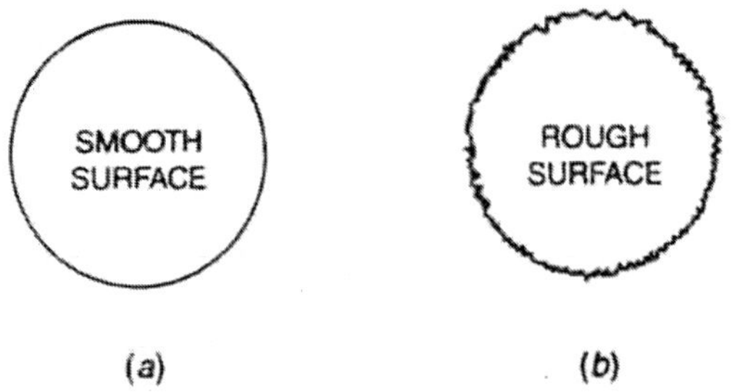

Fig. 3.16. Micrograph of the surface of tungsten wire.

3. **Facet Formations.** Facets are 'flat surfaces' found generally on the outer faces of the coil facing the glass shell. These do not change the energy balance of the filament with high working temperature. Facets play important role in low watt auto lamps with d.c. supply. The facets cause the low watt filaments to change their round shape to a hexagonal shape of a logenze and the coil breaks at these unbalanced angular spots. *[See Photo 3.4 and Photo 3.5]*

SOME IMPORTANT POINTS IN LAMP TECHNOLOGY

1. Operational datas for clear shell 230 volts incandescent lamp. Average life 1000 hours, lamp burnt on base up position.

Watt	*Lumen Output (Average through life)*		*Working Efficiency (Measured in lumens per watt)*	
	Single coil	*Coiled coil*	*Single coil*	*Coiled coil*
25	206		8.24	
40	330	389	8.25	9.27
60 *	584	665*	9.73	11.1* The economic lamps—
100 *	1,160	1,270*	11.6	12.7* For domestic use
150	1,970		13.13	
200	2,725		13.62	
300	4,430		14.77	
500	7,930		15.86	
1000	17,800		17.8	
1500	28,380		19.59	

The higher wattage lamps give more lumens per watt. The reason being that the loss of heat due to convection is proportional to the surface area of the filament and the current rating is proportional to the cross-sectional area of the filament.

Let us take the length of the filament to be 'L' and it's radius as 'r'. Then the complete surface area of the filament will be $2\pi rL$.

So, convection loss will be proportional to $2\pi rL$, but current rating (meaning lumens per watt as efficiency) is proportional to the small cross-sectional area of the filament which is πr^2. For higher wattage lamps when the length of the filament is increased, the heat loss by convection is not so high and, we get better efficiency for the lamp. For this reason, gas filling in lower wattage lamps as 15 and 25 watt is very uneconomical.

After setting for the wattage efficiency, let us now look into the part played by 'voltage' in lamps.

As in auto lamps, the 50 watt is very high for it's constructional restrictions and supply system.

But here the low pressure of say 6 or 12 volt to produce 50 watt can be utilised to advantage by the same πr^2 formula. For increasing the wattage here with low voltage, the diameter, of the auto lamp filaments are made very large which gives it sturdiness and long life. The increase in diameter here has increased the efficiency of the auto lamps.

2. The light emission from a lamp in proportional to the eleventh power of the absolute temperature, so that a very slight increase of temperature above the normally used in incandescent lamps, brings about a very considerable increase in efficiency.

3. The losses from the filament are of three types : first the direct radiation; second the conduction through the atmosphere and third the convection.

Direct radiation of the energy from the filament is desirable, since it is the radiation or a portion of it which constitutes the light.

The other two losses are harmful and should be minimised as far as possible; in commercial practice they have been reduced to the lowest limit by use of a very high vacuum.

4. Langmuir found that a filament formed into a closely coiled helix behaved, in so far as heat losses are concerned, much like a solid filament with a diameter equal to the outside diameter of the coil.

5. The current taken by a filament increases approximately with the $\frac{3}{2}$ power of it's diameter.

Diameter		*Current*
Inches	*mm*	*Amps*
0.005	0.127	3.0
0.016	0.254	8.5
0.20	0.508	24.0

Unless very low voltage are used the power consumed with the larger wires is so great that only very highly luminous lamps as auto head lamps can be made.

6. For *GLS* lamps, it was of vital importance to increase the effective diameter of the filament without decreasing it's resistance which was obtained by using a tubular filament—meaning winding a thin filament into the form of a helix—a coil.

7. About 20% of the energy radiated from the filament is intercepted by the glass which causes heating of the lamp. The convection currents carring the relatively large amount of

heat travel vertically upwards from the filament and strike a relatively small area of the bulb which causes it to overheat.

It is thus desirable, in small bulbs the filament should be placed in the lower part of the shell so that the heat would travel a longer distance and the deposits would have a larger area to spread over.

For the same reason, the bulb height is always considerably greater than the shell diameter. In *GLS* lamp, the filament has to be placed at the centre line of the shell.

Cooling effects of the leads lowers the efficiency of the lamps by an amount that is inversely proportional to the voltage and independent of the size of the wire or the current. For lamps running with less than 10 volts, it has a very serious effect.

8. The life of a filament is determined by the loss of tungsten by evaporation and has been found to be dependent on relative decrease in diameter caused by evaporation.

The evaporation of tungsten in argon—nitrogen atmosphere is a diffusion process. The rate of evaporation per unit surface area of the filament is approximately inversely proportional to its diameter. In very thin filament wires, the evaporation rate is nearly proportional to the squares of their diameter.

REFERENCES

1. Irving Langmuir—Physical Review—34,401, 1912.
2. D.A.Clarke—Modern Electric Lamps—
Blackie & Son Ltd., London.
3. W. Geiss—The Development of Coiled Coil Lamp—
Philips Technical Review Vol. 1/No. 4 ; 97–101, 1936.
4. E. Fisher and J. Fitzerald
Heat and Mass Transport in Cylindrical Gas—Filled Incandescent Lamps
Journal of Applied Physics 45/7 July1974.
5. H. Horster, E. Kauer and W. Lechner—
The Burn-out Mechanism of Incandescent Lamps
Philips Technical Review 32,155–164, 1971.
6. Collected Works of Irving Langmuir Vol. 2
Pergamon, Oxford, 1960.
7. W. Lews, H. Kinkartz and W. Lechner
The Influence of Facet Formation on the Life of Incandescent Lamps
Phil. Res. Repts. 30 : 218.225, 1975.

Photo 3.1. Irving Langmuir—Research Scientist. (General Electric Research Laboratory—1909–1957) Noble Prize winner 1932.

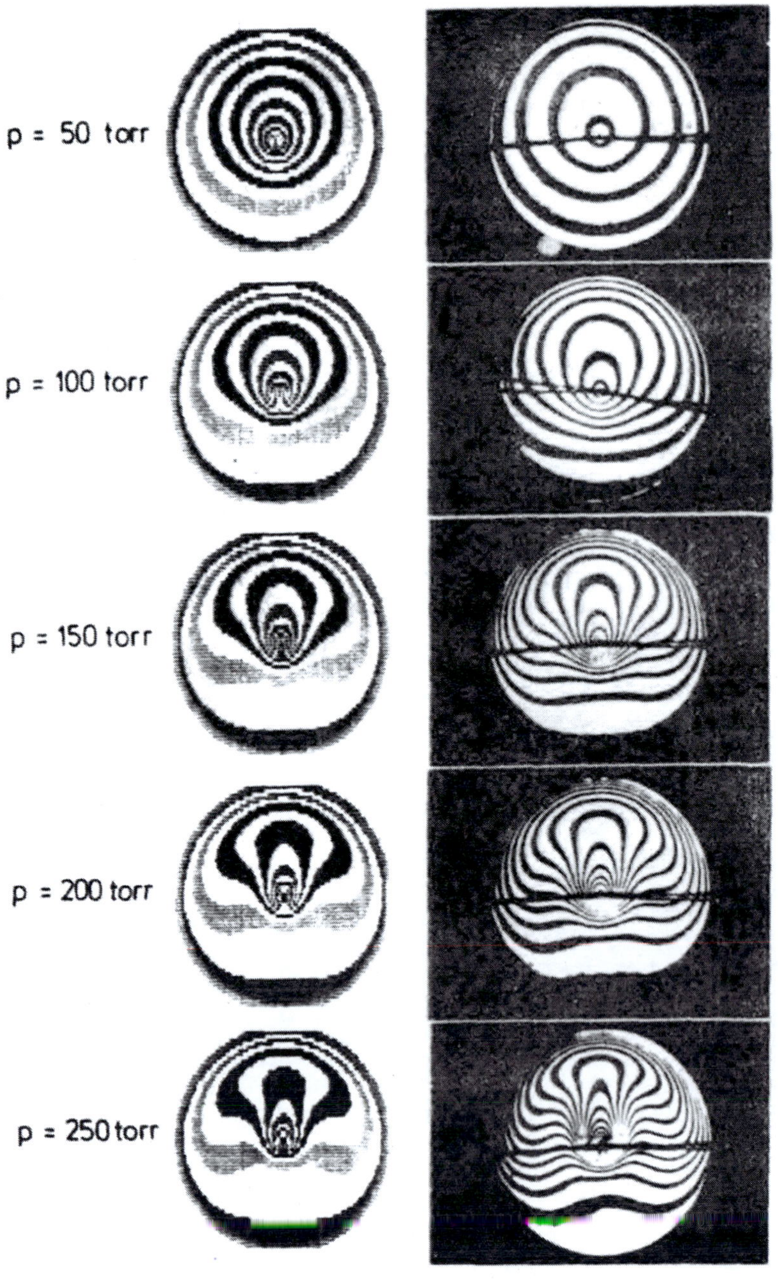

Photo 3.2. Holographic arrangement for temperature measurement. Two holograms of the lamp (taken before and during operation) are recorded on the same photographic plate. In the reconstruction process both situations are simultaneously reconstructed and differences in optical path appear in the form of interference fringes.

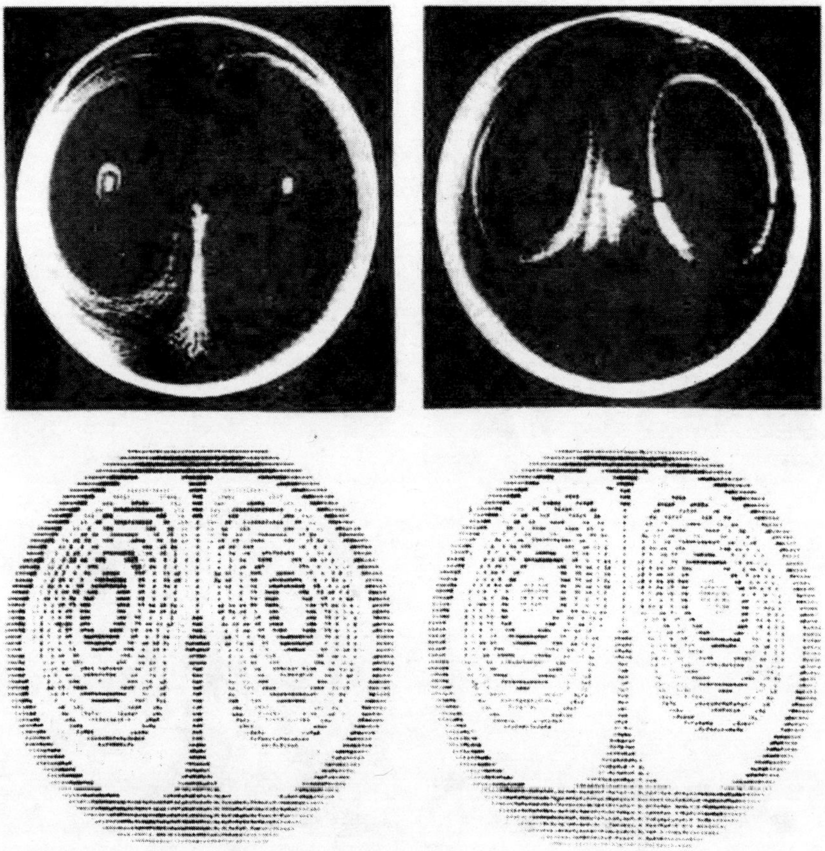

P = 200 torr
(a)

P = 300 torr
(b)

Photo 3.3. (*a*) Experimental arrangement for streamline measurement. Small light tracer particles are gently blown into the laser-illuminated test section : The motion of the gas is determined by recording the particle trajectories photographically.
(Courtesy : Philips Tech. Rev. 1795, Holland.)

Photo 3.4

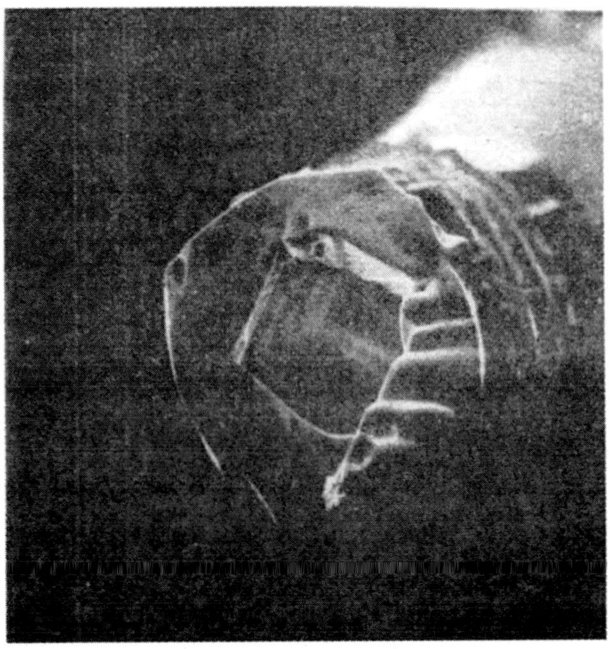

Photo 3.5

4

MEASUREMENT OF LIGHT

In the earlier chapter, it has been established that light is an electromagnetic radiation and how it causes sensation to the retina of the eye. Generally 'radiant power' or 'radiant flux' is expressed in watts. The word "flux" means continuous fluid state of any substance. The small band of radiant power between 0.4 to 0.7 micron which we commonly term as light has its own parameters of measurement. Most of our daily used measuring units for length, mass and time are independent of the involvement of the person measuring them. But for measurement of light, the units used are not purely physical but dependent on the sensation produced on an individual person which might vary from one another. Efforts are now being made to replace these human observations and evaluations with electronic and pyrometric impulses to avoid human error in judgements.

Human Factors in Lighting

The physiological factors of lighting change with age. The requirement of lighting increases with age. The following gives an approximation of the amount of lighting requirement for good vision.

Advancement of age	*Visual requirements*
10 yrs	1 (taken as unit)
20 yrs	1.5 times
30 yrs	2 times
40 yrs	3 times
50 yrs	6 times
60 yrs	15 times

Importance of Good Lighting

There are 4 principal criteria which are considered to provide good illumination :

1. Amount of light (lux)
2. Balance of lighting (Contrast)
3. Glare control
4. Colour rendering.

All these above characteristics of light are combined together to give a human being visual comfort, good visual performance and maintenance of healthy eyesight.

For measurement of light, the following glossary of terms are used in illumination and photometry.

Light. It is the 'radiant flux' capable of stimulating the eye to produce visual sensation. We know that the radiant flux as in case of any electromagnetic radiation is a physical quantity and it can be directly measured. In the case of light, it must be clearly understood that "light is the ability of the radiation to influence the physiological and psychological process of seeing through the power of very small clinical and electrical changes in the retina and brain". Naturally this conversion of radiant flux to visible light is a different function and very much dependent on human judgements.

Luminous Flux. This is the quantity of radiant flux needed to produce visual sensation and is measured in lumens. Its symbol is F or ϕ (Greek letter Phi).

Illumination. The amount of light falling on each square foot of a surface is called the illumination of the surface and is measured in "lumen per square foot", inlux or milliphot. It's symbol is E.

Brightness. We see an object through its brightness. This object might receive light from other source and seen by us through it reflection or this object may be itself be a light emitting source. This should never by misunderstood with "glare" which is discomfortable or disability condition of the eye caused by bright light.

The brightness of an object is measured for its two characteristics :

1. Objective brightness is known as luminance which depends solely on objective phenomenon and can be physically measured or created.

2. Subjective brightness is known as luminosity of an object which is simply the sensation of feeling of the quality of the light. This is dependent on environmental and psychological phenomena. For example a surface emitting twice the luminance from another surface (measured by a photometer) will not be shining twice as bright.

So the use of words luminance and luminosity always carry different meaning.

Also the difference between the two words *i.e.,* illumination and brightness are also very often misunderstood.

Illumination is the measure of the light received by any surface or the light fluxes falling on an object; whereas the object reflecting that light or emitting its own light is known as brightness. Illumination of an object is never equal to the brightness of the object. They are two different phenomenon. Whereas we measure illumination in lumens per sq. foot and brightness we measure in candela per sq. foot.

Luminous Intensity. It states the strength of the source or the ability to illuminate an object. This is the resultant of the radiant flux leaving the source and divided by the solid angle produced at the source while leaving. Here we must note two important factors as solid angle and steredian.

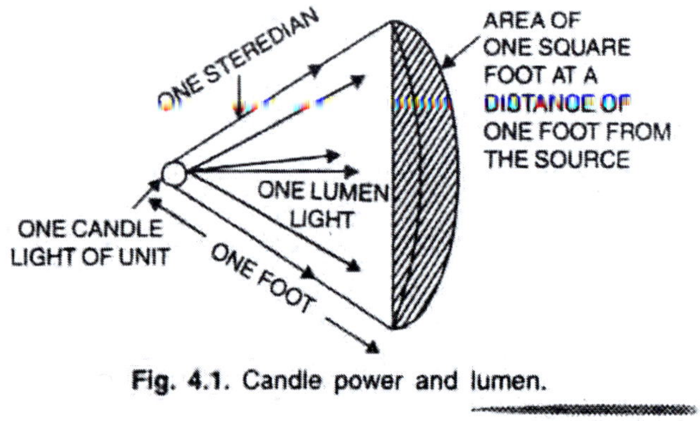

Fig. 4.1. Candle power and lumen.

Consider a portion of the curved surface of a sphere. The radius of this sphere is a unit say 1 cm or 1 foot. The curved surface we take as a square of the unit as 1 cm^2 or 1 ft^2. Now the angle subtended by this surface as cone is known as a "solid angle". When we have taken this curved surface as square of the unit, it is called a 'steredian'.

Since the area of the surface of sphere is 4π, so the total solid angle produced at the centre is 4π steredians. For calculations of illumination and study of photometric integrators, these two terms are to be properly understood.

Candela. It is the unit of luminous intensity. This is the fundamental unit for measuring luminance. Instead of relying on human judgements, universal physical factors have been applied to set this unit.

This is the luminance of a full black body **radiator** at the temperature of solidification of platinum taken to be as 60 units, of luminous intensity per square centimetre. (While describing about standards, the black body radiator is described in detail).

Candle Power. The light radiating capacity in a given direction when the luminous intensity is expressed in candelas.

Mean Spherical Candle Power (MSCP). The average value of candle power in all direction. For calculating total Lumen this is used.

$$\text{Total lumen} = 4\pi \times \text{MSCP lumen.}$$

Lumen. This is the unit for measuring luminous flux. Lumen is not a measure of the sensation (as it is often erroneously understood) but it is a measure of the stimulus. For common use, luminous flux is expressed in lumen and the other photometric quantities in lumen per sq cm or sq ft. etc. The quantity of flux emitted by a source of light having an uniform intensity of one candela at a unit solid angle of one steredian is taken as one lumen.

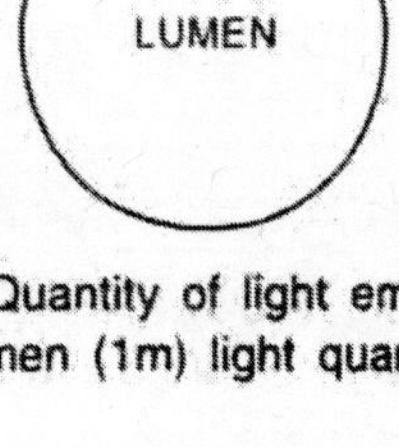

(*a*) Quantity of light emitted
Lumen (1m) light quantity

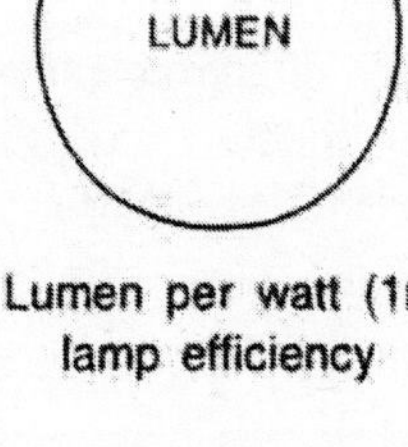

(*b*) Lumen per watt (1m/W)
lamp efficiency

Fig. 4.2

A lamp whose candle power is one candela is said to give out one **lumen** in one solid angle of 1 steredian.

Calculation of lumen being the most important issue in measurement of light, the same is being explained in the following observations :

Relations between Candles, Lumens and Foot Candles. A uniform point source of one candle power is shown at the centre of a sphere of 1 foot radius. It is assumed that the sphere is perfectly transparent. The solid angle subtended by the area *A, B, C, D* is 1 steredian. The flux density is therefore 1 lumen per steredian which corresponds to a luminous intensity of 1 candle power as originally assumed.

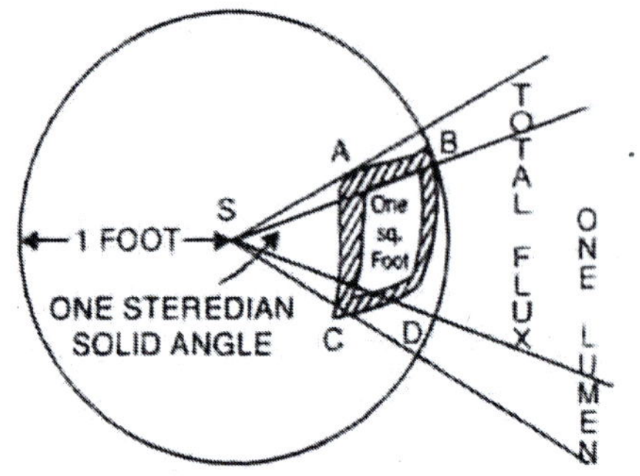

Fig. 4.3

The sphere has a total area of 12.57 sq ft. and there is a luminous flux of 1 lumen falling on each square foot. Thus the source of 1 candle power produces 12.57 lumen.

From the above figure we see that the source of the light is a unit of one candela. This is placed at the centre of the sphere having one foot radius. An area of one square foot on the sphere subtends one steredian solid angle at the source. So the quantity of flux emitted is one lumen.

Solid angle around any point is 4π steredians so it will be with the above source "S" also. The total 'lumino is flux' emitted by the source *S* is 4π lumens.

We know π is $22/7 = 3.14$ so $4 = 4 \times 3.14 = 12.57$. lumen. It would mean that a lamp of one candela emits a total 12.57 lumen of luminous flux around it. Also it is seen that the lumens emitted in unit solid angle is numerically equal to the candela in that direction.

For practical purposes, most of the lamps cannot give out same flux equally on all directions around the lamp. An ordinary naked bulb will have a low light value when seen from the cap side as the luminous fluxes cannot travel through the cap. Whatever fluxes go up, generally do not serve useful purpose; so various types of light dispersive fittings are used to reflect the upward going fluxes to useful direction. It would be worth while here to know about the usage of "polar curves" of lamps and how they help in adjusting illumination levels.

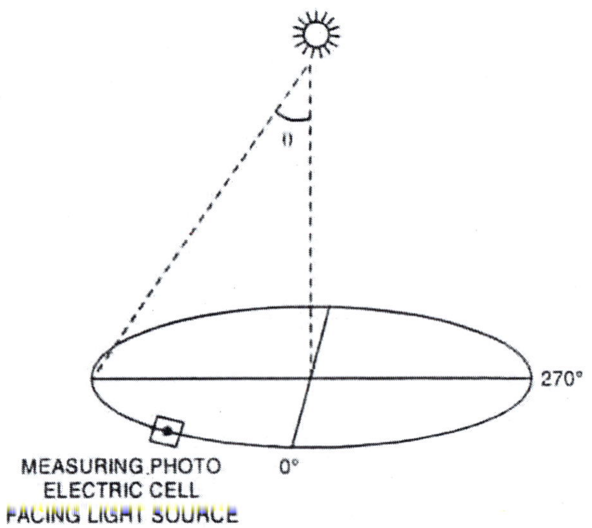

Fig. 4.4

The following "polar curves" are of a 100 watt lamp. These gives the strength of the light as candela at different directions from the lamp. The 0° to 180° curve is drawn from the bottom through the centre of the cap vertically to the top. The rest of the degrees follow as in the drawing. The filament is kept horizontally parallel with the ground. For a naked light, candle power or candela is reduced near the cap; also the light in the horizontal plane is less than the vertical plane.

The thick curve of a naked lamp proves that *OA* which is 20° from the vertical gives 50 candelas whereas *OB* which is 90° with the vertical gives 45 candelas. The dotted curve of the shaded lamp has completely changed the illumination positions. For every lamp shade and fitting, the manufactures supply polar curves showing candle power distribution with the specified type of lamp.

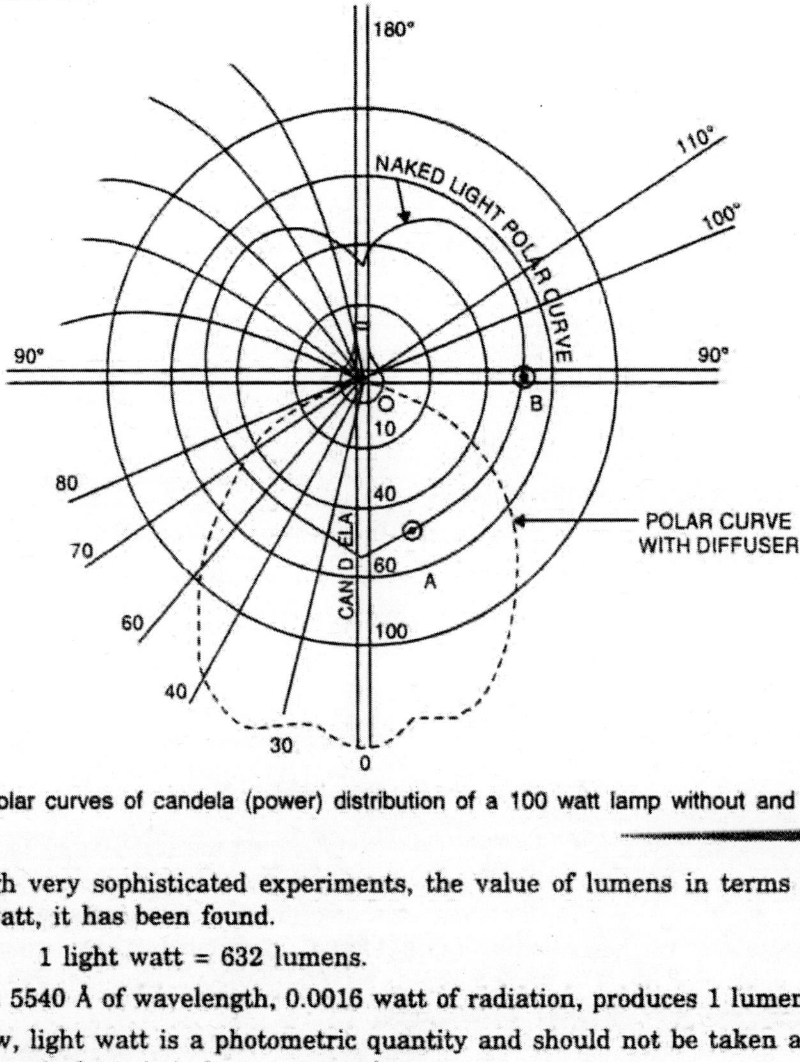

Fig. 4.5. Polar curves of candela (power) distribution of a 100 watt lamp without and with shade.

Through very sophisticated experiments, the value of lumens in terms of commonly used unit watt, it has been found.

1 light watt = 632 lumens.

Also at 5540 Å of wavelength, 0.0016 watt of radiation, produces 1 lumen of light lux.

Anyhow, light watt is a photometric quantity and should not be taken as a source of mechanical equivalent, it is just a comparison.

Lux. For measurement of light, this form will be very often heard as our country has adopted metric system.

This is a unit of illumination. An illumination of one lumen per square metre is called a Lux or Lx.

1 Lux = 0.0929 lumens per sq foot

1 lumen per sq foot (or 1 foot candle) = 10.76 Lux.

Nit. This is a unit of luminance as one candela per square metre.

Stilb. Luminance of one candela per sq cm is one stilb or sb.

Luminance Conversion Table

Terms	cd/m^2	cd/cm^2	cd/ft^2	*ft*
Candelas per sq m (nits)	1	0.0001	0.0929	0.2919
Candelas per sq cm (Stilb)	10.000	1	929	2919
Candelas per sq ft	10.76	0.001076	1	3.1416
Foot—lamberts (ft L) for measurement of brightness	3.426	0.0003424	0.3183	1

Some important symbols and abbreviations used in Illuminating Engineering,

μ (micron)— wavelength

mμ (millimicron)— wavelength

λ (Lambda)— wavelength

ν (nu)— frequency

ρ (rho)— reflection factor

τ (tau)— transmission factor

F or ϕ (Phi)— luminous flux unit : lumen (lm)

I— luminous intensity unit : candela (cd)

E— illumination unit : lumen/sq.m. (lx)

Ω, ω (omega)— solid angle

Δ, δ (delta)— a very small solid angle.

LAWS OF ILLUMINATION

1. **The Inverse Square Law.** This very common and natural phenomenon of light is mostly misunderstood even by well informed people. Generally the light emitting quality of a lamp is criticised while a lamp is being at a distance from observers. Everyone speaks about their feeling of the sensation of light. No one thinks about their distance from the source of the light. Specially the setting of auto head lamp focus when not done at a correct distance, proper results can never be obtained.

Distance plays a great part in the strength of light. The observer may not be conscious of a difference of one or two feet while feeling the effect of light but actually it affects the strength inversely to the square of the distance.

The inverse square law states that—the illumination of a surface normal to the incident light is inversely proportional to the square of the distance of the surface from the source.

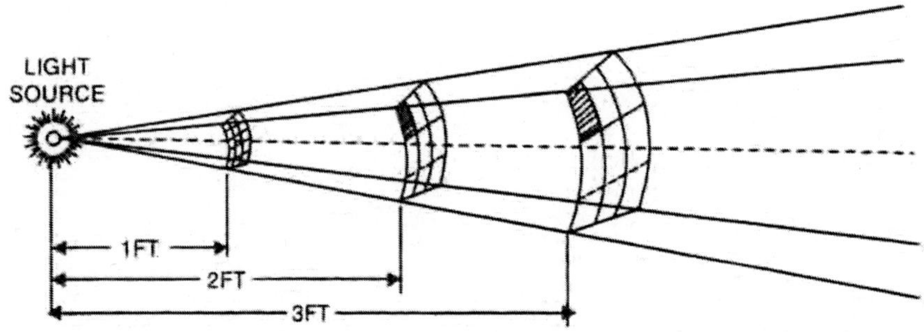

Fig. 4.6. The inverse square law of illumination.

$$E \text{ (illumination)} = \frac{I\text{(candela)}}{D^2\text{(distance from source)}}$$

$$\text{Illumination } (E) \text{ at 1 ft} = \frac{1}{1^2} = \frac{1}{1} = 1 \text{ lumen or candela}$$

$$2 \text{ ft} = \frac{1}{2^2} = \frac{1}{4} \text{ candela}$$

$$3 \text{ ft} = \frac{1}{3^2} = \frac{1}{9} \text{ candela}$$

In bunsen grease spot photometer, lamps are compared on a graduated scale. Thus difference of 1 or 2 cm between two lamps would mean a great anomaly and actually show the different lumen power of the two lamps.

2. **Cosine Law of Illumination.** This is another fundamental law of illumination which is intermingled with the Inverse square law. Actually for any calculation purpose, the combined laws of inverse square and cosine attributes are used.

The cosine law states—if the illuminated surface is tilted, the luminous flux falling on unit area is reduced in proportion to the cosine of the angle (theta) between the incoming ray and the normal to the surface.

Combining the two laws we can now calculate the illumination of any surface by any source (of course it should not be only too close to the surface.)

$$E \text{ (illumination)} = \frac{I\text{(luminous intensity candle power of the lamp} \times \cos\theta)}{d^2\text{(distance from the source)}}$$

With the help of the above formula, the illumination of any place can be easily calculated. We have to obtain the value of the cosine of the angle from the following chart.

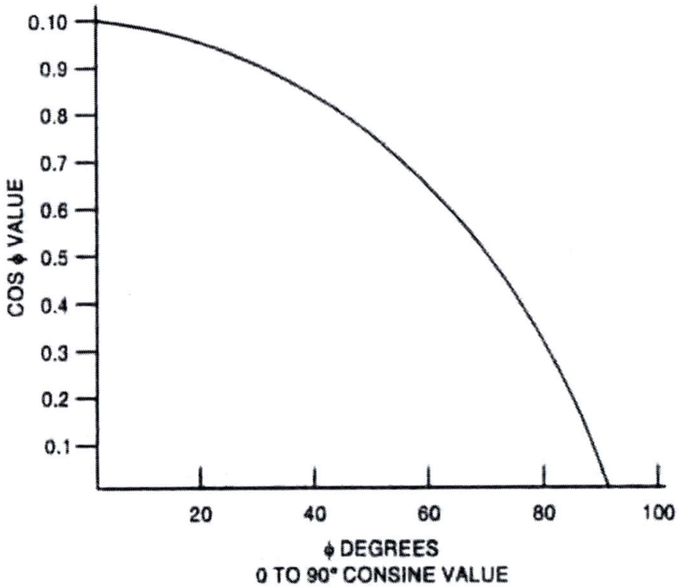

Fig. 4.7

Example. A shaded lamp is hanging 12′ away from the work bench and the light from the lamp strikes the bench at 40° from the right angle to the surface of the bench (known as 40° from normal). If the lamp shade manufacturer states the candle power of the lamp to be 900 lumen in the direction of the bench, then calculate the illumination on the work bench.

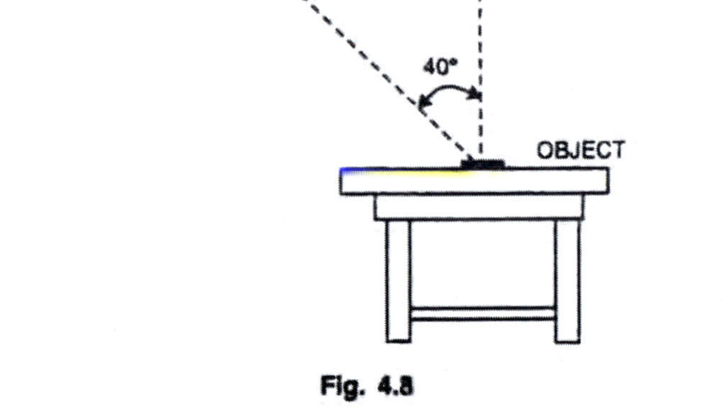

Fig. 4.8

Formula.

$$E = \frac{I \times \cos\theta}{d^2}$$

$$I = 900 \text{ lumen}$$

$$\cos 40° = 0.75 \text{ (from chart)}$$

$$E = \frac{900 \times 0.75}{12' \times 12'} = 4.69 \text{ lumen/sq ft.}$$

So we can now summarise that the illumination produced by a lamp on a surface depends on the following :

1. The luminous intensity of the source (I).
2. The distance between the surface and the lamp.
   (Inverse square law)
3. The angle at which the light rays strike the surface.
   (Cosine law).

STANDARD

Whenever we measure any entity, we express our result in a numerical value. The numerical value has a unit which has been established, accepted and maintained by a country or the world as a whole. This unit relates to a specially maintained object or system which remains unaltered due to time or weather.

Unlike physical measurements, light is concerned with psycho-physical response to a very-very narrow band of electro-magnetic radiation perceived by human being as light. Man can only express variations or equality and build instruments in comparison to his perception. After this stage of instrumentation only the light then can be numerically expressed.

The very first thing in light is to describe the luminance meaning the intensity of the light—how powerful it is—with what it is to be compared with—how a standard is to be set. For all practical purposes, in the early days of optics light was standardised with the power of a candle. A standard candle is one which is made of sperm wax weighing 1/6 pound, 7/8 inch in diameter and burns at the rate of 120 grains per hour. For finer measurements, it is very difficult to maintain the purity and specifications of a sperm candle.

A standard is called as such because this is reproduceable from specifications. So, to make studies and set standards in radiation, the black body principle has been applied. A black body is a body that absorbs all the radiation that falls on it. At the same time a black body also radiates more power in one wavelength than any other type of radiating body. So, it was decided that if a piece of pure metal could be melted, the luminance produced by it could be taken as a standard as it qualified for the following requirements :

(*a*) The specifications for the metal could be defined, settled and repeated whenever necessary.

(*b*) The procedures for melting would be recorded and repeated.

(*c*) The result could be verified with repeated performance.

(*d*) Once all the results are recorded and passed after retesting them, this would become a standard which could be repeated anywhere in the world.

So, for setting up primary standard of light, a full radiator is operated at the temperature of 'solidification of platinum'.

The Black Body Radiator

The construction of a black body radiator to obtain photometric standards is given below.

The complete radiator consists of a fused thorium oxide tube having 2.5 mm internal diameter and of about 45 mm length. This is known as the 'black body' proper. The tube is placed centrally in a crucible made of the same fused thorium oxide and the space between the tube and the crucible is filled with pure grade platinum. The top cover of the crucible locates the central tube and a small hole at the centre is provided for the radiation to emerge outside. The crucible is again embeded in a fused and unfused thorium as shown in the figure. The complete unit is wired by high frequency induction coil to melt the platinum inside the black body. The complete furnace alongwith its content are heated till the platinum melts completely. (Owing to multiple reflections, the strength of the radiations is increased and is visible at the small opening). With the platinum in molten state the furnace is switched off and the platinum starts cooling slowly. The temperature drops till 1773°C when platinum inside starts solidifying. This is known as the 'utectic stage of' of a metal when temperature remains constant but physical changes take place internally.

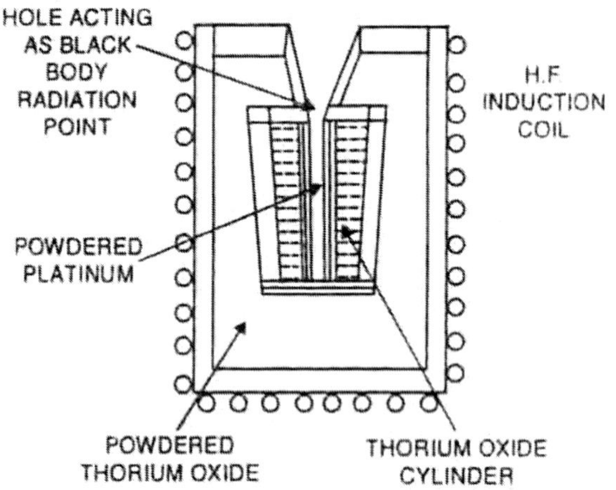

Fig. 4.9

The temperature at this stage for platinum remains constant at 1773°C for about 20 minutes. This is the most ideal period when the radiations will be constant and many readings could be made of the luminance it produces. During this period of steady and constant temperature the luminance has been found or accepted as 60 candela per square centimetre. This is the standard of luminance in photometry. Now electronic devices record this ever repeatable physical phenomenon and convert it to a known quantity.

This is the primary or original standard in photometry. Once the primary standard is set the other standards can be easily worked out.

Setting of Standards in Electric Lamp

The primary standard is a sophisticated more elaborate and time consuming process which is carried out very occasionally and special lamps are made as for the particular functional requirement. This is generally kept in the National Laboratories and at Standards Setting Institutions. So for working purposes the standards are set in the following manner.

Primary standard, secondary standards, working standards.

Secondary standard lamps are made in comparison with the primary standard lamp. The secondary standard lamp generally does not have the same physical shape of an actual working lamp. This is also treated as a sophisticated item for daily use. The secondary lamps is of robust construction with larger glass shell so that it has more area to absorb the vaporized filament. The filament is made in one plane and is welded to the lead-in-wires. For this lamp the main criteria is to maintain constant luminance for a long period of time. The secondary standard lamp should not be aged less than 100 hours before being put to service.

Working standard lamps are made by the individual manufacturers in their own quality control laboratory in comparison with the secondary standard lamp for making 'spot checks' in daily routine work. The working standard lamp is of the same shape as normal lamps in the production line. Special care is taken from the initial stages of production so that these attain the best of technical performance. By use of working standard lamp, the working load and risk of breakage and early failure of the secondary standard lamp is reduced. This is a common practice but for actual and perfect measurements, the service of the Secondary Standard Lamp has got to be utilised.

(See Photo 4.1. Reference Standard Lamps)

Photometry

The word photo means light and metry means measurement. After manufacturing a lamp the most desired qualities of its output, *i.e.*, the light has got to be measured for which the help of photometry is sought.

In photometry the following types of apparatus are in use to measure different characteristics of light.

Spectroscope. This is used for separating light into different components of colours.

Spectrometer. This is used when different wavelengths are to be measured.

Spectrophotometer. When an apparatus is used to measure visual radiant energy at various wavelengths, it is known as spectrophotometer.

Illumination is affected by the following four factors and can be measured by the types of photometers as mentioned alongwith them.

1. Luminous intensity of a source—test it with Bunsen Grease Spot Photometer.

2. Luminous flux from a source—test it with Photometric Integrator.

3. Illumination of a surface—test it with Macbeth Illuminator.

4. Luminance resulting from illumination—test with Lummer Brodhun Luminance Meter.

For testing of lamps, the luminous flux and the luminous intensity are the two important factors. Again when the luminous flux is determined the intensity of the source can be calculated.

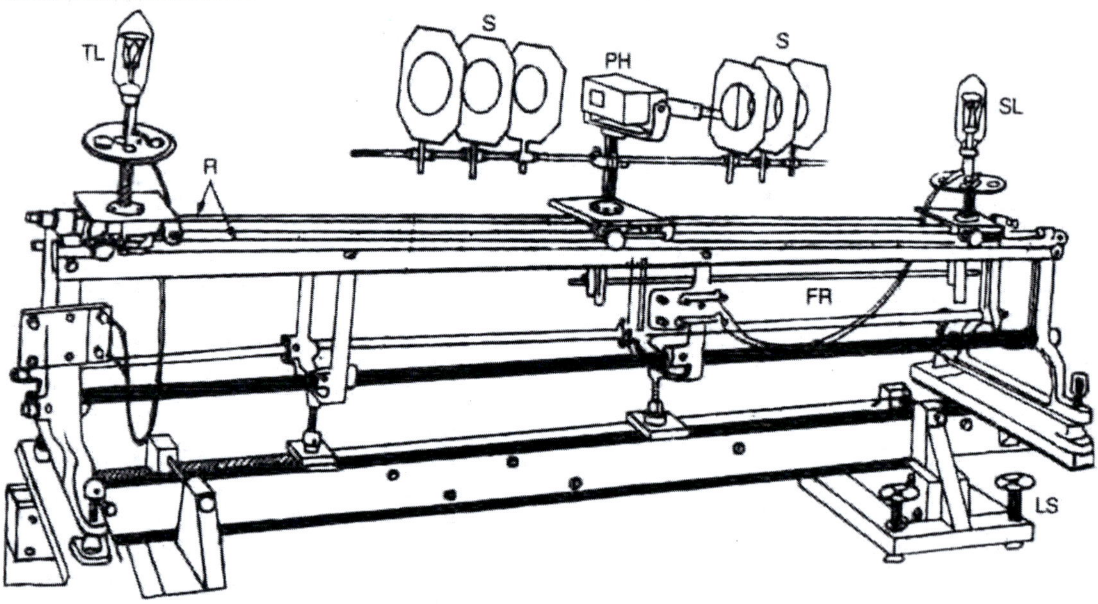

PHOTOMETER BENCH

Fig. 4.10. TL—Trial lamp, S—Screens, PH—Photometer head, SL—Standard lamp, R—Rails, FR—Distance fixing rod, LS—Levelling screw.

1. **Measurement of Luminous Intensity** (*Photo 4.2*). The easiest and cheapest way to measure the intensity of luminous flux could be carried out by a bunsen grease spot photometer. In the photographic view at 4.2 a lamp under test for its intensity of luminous flux is being compared with that of a working standard lamp.

The screen is simultaneously illuminated by both the working standard lamp and the lamp under test. The carriage holding the test lamp is moved horizontally till equal luminosity is obtained on to the screen. As per inverse square law if both the sources of illumination be of same strength, then

$$\frac{I_1}{d_1^2} = \frac{I_2}{d_2^2}$$

When we know the value of I_1 (Working standard lamp)

then $$I_2 = I_1 \times \frac{d_2^2}{d_1^2}$$

The grease-spot photometer has a screen made of white paper of zero transmittance at the middle of the rail. At the centre of the paper screen a very small spot of grease is placed. Though this spot looks dark, it transmits some light through it but it does not reflect any light. If the screen is illuminated from one side, say by the standard lamp, then this side of the screen looks bright with a dark spot at the centre. Now the 'lamp under test' is switched 'on' and the brightness of the central grease spot will change. By sliding the movable base of the 'test lamp' towards or away from the screen, a position will come when on both the sides of the screen the luminosity of the grease spot will equal the brightness of the base white paper.

The figure shows the arrangements of a photometer bench testing two secondary standard lamps. While testing a lamp from the production line, its base is to be adjusted vertically so that its filament centre length equals that of the standard lamp. When measuring lamps, the current and voltage of the lines are to be kept under strict control. It has been established that the lumens value of the lamp changes at six times higher rate than the change in the flow of current and this change is 3.6 times more than the rate of the change in the applied voltage. To achieve 99% accuracy in any type of photometric measurements the voltage and current are to be stabilized within 3 to 4 parts in thousand.

To overcome inaccuracies caused by stray lights such as reflection of the face of the observer and other objects, a series of thin blackened screens are placed on both the sides of the photometer head. Also black velvet screens are being provided around non-working sides of the lamps to avoid any disturbance from reflection.

2. **Measuring of Luminous Flux.** For tungsten filament lamps, the total luminous flux produced by them are of great importance. Most accurate measurement of luminous flux is done by goniophotometer. (See Photo 4.3)

The goniophotometer is operated in a totally dark room. The earlier types, the operating distance for optical observation was 7.3 metres which is now reduced to half in newer models by using a mirror. The lamp is bolted to an arm which can be moved vertically and around its axis as well. The light centre is set to the axis of the rotation of the mirror by measuring its distance down the support bar and then transferring this figure to an indicator on the gantry which stops at any desired position when raised. Thus a goniophotometer allows precise adjustment of angle in both elevation and azimuth. It is also specified that the optical path for the flood light should not be less than 30 mtrs and that for auto head lamps 25 mtrs.

With this photometer a series of readings are taken for the lamp at different distances and different angles. The readings give the luminous intensity of the lamp in all directions in the space. When the average value of these readings is multiplied by 4π the most accurate value of the luminous flux is obtained.

Photometric Integrator. Photometric integrator is described as an instrument which enables luminous flux being emitted by a source to be determined by a single measurement.

A photometric integrator incorporates measurements being taken in a hollow sphere which had been invented by Herr Ulbright of W. Germany. The theory behind it is that the internal sides of a perfectly hollow sphere having a matt white diffusing surface with equal reflecting factor at all points when lighted with a lamp any where inside it will always show luminance, on the internal surface proportional to the total luminous flux of the lamp.

The internal surface of the integrator must be perfect with a true centre. For GLS lamps, the internal diameter of the integrator should be at least six times more than the length of the test lamp without the cap. For most accurate measurements for flood light test, the diameter of the sphere is sometimes 5 to 6 feet.

The facilitate photometric measurements, a small circular window is provided. This window should not be more than one millimetre above or below the internal surface of the integrator. The diameter of the window should not be more than one tenth of the diameter of the integrator and its internal surface also is to be painted with the same paint like rest of the surface.

Normally two to four coats of finely precipitated barium sulphate dispersed in water with fibre free sodium carboxymethyl cellulose is sprayed. As consistancy is the main requisite, any sign of yellowing would necessitate in washing and repainting of the integrator.

The arrangements inside a photometric integrator is shown in the following figure. (*Also see Photo 4.4 and Photo 4.5*).

I — Integrator
P — Photo cell
ID — Iris diaphragm
W — Window
L — Lamp under test
S — Screen
A — Auxiliary lamp
AS — Aux. lamp screen.

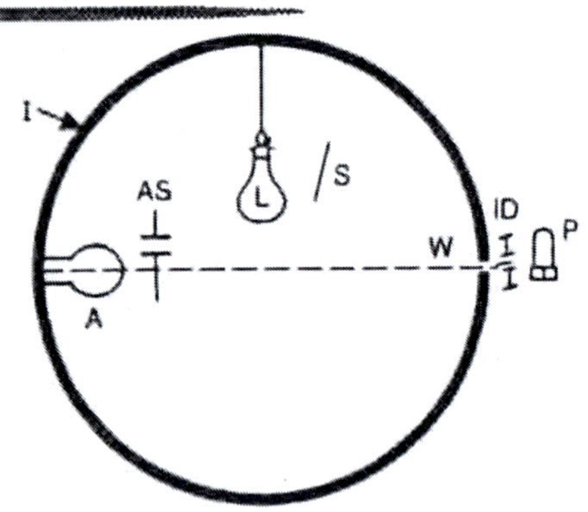

Fig. 4.11

In the figure, *I* is an integrator which may be made of steel copper, zinc, aluminium or plaster of Paris. The internal arrangements are to be made as the situation calls for. *L* is the lamp to be tested. To avoid light falling on the diffused window *W*, the screen *S* is interposed between them at one third distance of the lamp and the window. A lamp within an integrator will absorb a portion of its own luminance flux. The readings will not vary if these are obtained from a lamp which is of the same size, shape and type as the standard lamp. If lamp of other size, shape and type is used, an auxiliary lamp *A* is placed near the internal wall of the integrator and a screen *AS* is placed before it for the light to fall from it to the window or on the test or standard lamp.

First a reading is taken with the auxiliary lamp, then the standard lamp is placed at *L* but not switched on. One more reading is taken in this condition. The difference in the two readings are noted. Now the standard lamp is replaced with the lamp to be tested and one more reading is taken. The ratio of the standard lamp and the test lamp will give the corrective factor.

In a factory, lamps of different luminous intensities are to be tested; so it is useful to install an Iris diaphragm to control the amount of flux falling on the photo cell. Also it is very important to install the photo cell in such a manner so that it remains fully illuminated at all times. When Iris diaphragm is used the readings in the micro-amperemeters used for measuring the current could be used directly without going into calculations for illumination factors.

In the case of testing linear type of lamps as Halogen lamp, this is to be placed at right angle to the line passing through the centre of the integrator and the window interposed by a suitable screen.

The manufacturers of the integrators always state the special cares to be taken for operation and also supply with the conversion tables for different lamps. The integrator gives the value in lux and this can easily be converted into lumens or other units as stated in the conversion table.

TESTING OF INCANDESCENT LAMPS

Generally measurements and checks for the following points are made in a lamp :

1. Dimensions of the lamp—Check it with vernier callipers, micrometer, scales etc.
2. Light centre length—Centre length measuring gadget made with a scale.
3. Markings on the lamp—Check it as per Indian standards specifications.
4. Insulation resistance of the cap—Check it with a megger.
5. Bonding strength of the cap cement—Check it with a torsion tester.
6. Heat rise of the lamp cap—Check it with the thermocouple device.
7. Initial wattage of the lamp—Check it with voltmeter and ammeter.
8. Total luminous flux and lumens values—Check it with photometric integrators or bunsen grease spot photometer.
9. Total life of a lamp—Check it on life testing line.
10. Pre-focus angles and dipper angles of auto lamps—Check it on distant screens or walls.

Automatic Digital Photometer for Testing Incandescent Lamps. (*Photo 4.6*)

In an incandescent lamp factory regular photometric measurements have to be made on large numbers of lamps by way of sample inspection. Done by conventional methods these measurements call for constant attention of a high order from the operator. After placing the lamp in the photometer the operator must accurately adjust the voltage on the lamp to the required value by means of a meter reading (and keep it at that value). The operator must then carefully read the luminous flux and the power consumption from two metres, enter the readings for each lamp without error in the inspection sheet, divide one value by the other to find the luminous efficiency (lumens per watt), divide this in turn into the luminous-efficiency figure specified for the particular type of lamp, and again record both results. The process is time consuming and abhorts inaccuracy.

The equipment illustrated here, which is in use at Philips' new incandescent lamp factory at Weert (Netherland), performs practically all these operations automatically. The operator has no meters to read, no calculations to perform and very little to record. She selects the voltage on a keyboard at her right on the table (not visible in the photo) : there is a choice from 80 values between 90 and 287.5 V. An automatic control system keeps the lamps voltage at the selected value to within 0.1%. On the same keyboard she selects the luminous efficiency specified for the type of lamp under test and the appropriate (coupled) ranges on the watt and lumen meters (choice of six ranges). The selected values of voltage and specified luminous efficiency are automatically printed on the inspection sheet which the operator has inserted in the electric typewriter and on which, where necessary, she has typed a few identifying particulars of the type of lamp tested. After placing each lamp in the photometer (on the extreme left in the photograph) the operator presses a button. The system automatically allows the lamp to burn for a preset time, *e.g.*, 10 seconds, sufficient for it to reach a steady state, after which the system automatically carries out the measurements, performs the calculations mentioned and print out the results on the inspection sheet. The progress of these automatic operations is signalled by pilot lights on the keyboard.

The installation is composed of series-produced industrial equipment. The cabinet seen on the right contains in the left-hand part two automatic potentiometers for measuring the lumens and watts, one automatic potentiometer which acts as an analogue computer for the division operations, and above an analogue-digital converter, which translates the continuously variable results of the measurements and calculations into the discontinuously variable signals required for driving the electric typewriter. A fourth automatic potentiometer works as a servo-system for the final adjustment of the lamp voltage (which thus takes only a few seconds). The standard of accuracy required makes it necessary to operate the lamps from a DC supply. The DC generator used (not shown in the photograph) is fed by a stabilized power pack, bottom left in the cabinet.

The right-hand part of the cabinet contains the programme switch and relays for controlling the automatic measuring and computing cycle, and all switches for adjusting the voltage, meter ranges and so on. The switches are operated by remote control from the above mentioned push-buttons on the keyboard. Grouping them in one cabinet together with the appertaining meters made it possible to simplify the wiring. The equipment as a whole is checked twice a day by measurements on standard lamps.

An Example of Lighting Design for Special Purpose Illumination

(*See Photo 4.7. Nehru Stadium*)

The following lighting parameters were carried but by Philips India Ltd. at the 9th Asian Games at New Delhi which give an idea of the level of illumination needed for such indoor and outdoor activities.

Jawaharlal Nehru Stadium (*Photo 4.7*)

	Illuminance Levels		
	Stage I for training/ recreation	Stage II for black & white TV	Stage III for colour TV
(*a*) **Football Pitch**			
(*i*) Average vertical illuminance on football pitch at 1.5 metres above arena level, in the direction of main cameras.	Over 600 lux	Over 1400 lux	Over 2200 lux
(*ii*) Vertical uniformity ratio ($E_{min}/E_{average}$).	0.8	0.8	0.75
(*iii*) Average horizontal illuminance football pitch at zero level of arena.	Over 800 lux	Over 1900 lux	Over 2600 lux
(*iv*) Horizontal uniformity ratio ($E_{min}/E_{average}$).	0.9	0.8	0.85
(*b*) **Entire Arena including Tracks**			
(*i*) Average vertical illuminance for the entire arena including tracks.	Over 600 lux	Over 1250 lux	Over 2000 lux

Photo 4.1. Reference standard lamps—BIPM type.
(Courtesy : National Physical Laboratory, New Delhi).

Photo 4.2. Photometer bench.
(Courtesy : NPL, New Delhi).

Photo 4.4. Integrating sphere (Exterior).
(Courtesy : NPL, New Delhi).

Photo 4.3. Gonio photometer
(Courtesy : Crafts de Fluorescent, Bombay 78).

Photo 4.5. Integrating sphere (Interior).
(Courtesy : NPL, New Delhi).

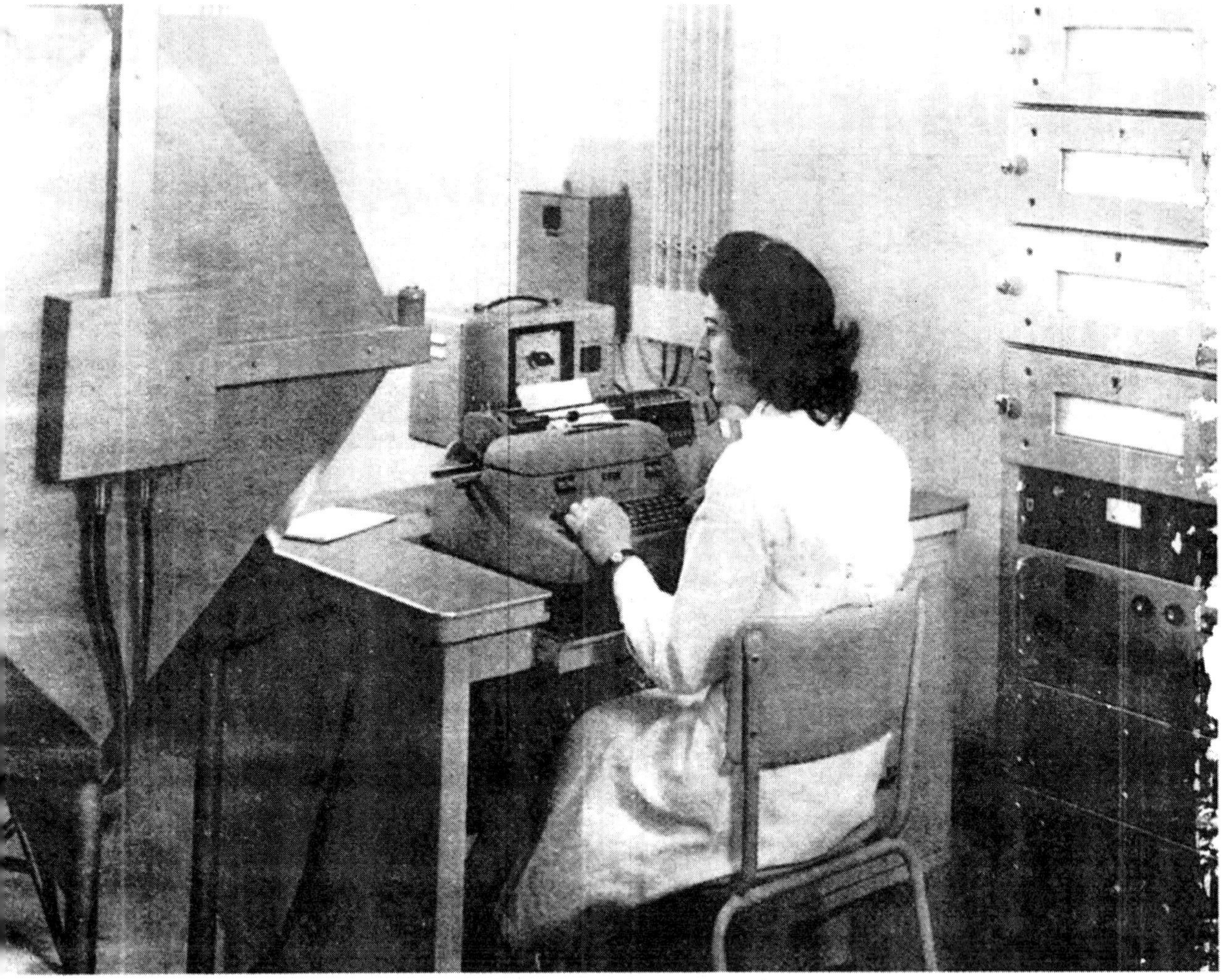

Photo 4.6. Automatic digital photometer for testing incandescent lamp. (Courtesy : Philips Tech. Rev, Holland).

Photo 4.7. Illumination at Jawaharlal Nehru stadium.
(Courtesy : Philips Professional Journal, India).

(*ii*) Average horizontal illuminance for the entire arena including tracks.	Over 750 lux	Over 1700 lux	Over 2400 lux

(*c*) **Gradient for Entire Arena**

10% or less over a distance of any 4 metres.

Hall of Sports, Mathura Road

Average horizontal illuminance : Over 1000 lux

Average vertical illuminance : Over 500 lux

Indoor Stadium at Talkatora Gardens

	Average Horizontal Illuminance
Basketball court	Over 1600 lux
Entire arena for other games	Over 1400 lux
Entire arena for conference	Over 550 lux
Entire arena for emergency	Over 350 lux

All values are initial, and to arrive at maintained values, maintenance factor is 0.8. *Courtesy*. Philips Professional Journal, India.

Some Do and Don'ts for Lamp Testing

All measurements and tests are to be carried out as per the standards set out by the Indian Standards Institution. The requirements specified in the standards include those for mechanical and physical characteristics, specifically in respect of glass shell, cap and holder, lumen output, lumen maintenance and life for lamps of the three varieties as Mini, Auto and GLS. The standard not only facilitates interchangeability but also ensures the availability of quality lamps to the consumers as advanced by the latest manufacturing technology which is advocated by the international electrotechnical commission.

While setting up a testing laboratory for electric lamps, utmost care is to be given to the supply voltage. Testing of electric lamp is a sophisticate and very accurate job. All tests are carried out by electronic equipments. If the supply voltage itself is erratic, the results would be just a bunch of meaningless figures. So, simple type of voltage stabilizers are used for other electrical types of electrical equipments are not suitable for lamp testing. For lamp laboratories, servo type stabilizers providing less than 1% voltage fluctuation must be used. This would be needed for use with photometric integrator, bunsen photometer, initial ratings, testing and life testing. For life test the mean effective voltage is to be obtained by a V^2 meter as described in the GLS lamp specification standard.

It is considered that proper assessment of the qualities *i.e.*, 'life' and 'light' of a lamp cannot be made by burning it in excess of rated voltage as the complete science of manufacture is based on those standards. The results so far obtained and accepted by us are all made in the most advanced countries where the filaments, electrodes and the process of manufacture are mostly carried out through latest technology. So, forced burning of lamps in our country may not provide us with a proper guidance to testing. Practically the filament and the electrodes fail very rapidly and most unexpectedly. Where time permits, this practice of forced burning should be avoided.

When the hermetic sealing is done properly, the gas leakage from the lamp will be a very slow affair. As a guideline for a good lamp, it can be said that a hundred watt lamp at its first hour of burning if maintains a steady 104 watt, can be termed as a sign for a good lamp. In most cases these types of lamps burn between 750 hours to 1000 hours and provide expected lumens. The first burning hour of the lamp is very critical and the temperature, light and voltage are to be very carefully checked at this period. The lamps giving more light would mean fitting of stretched filaments and their life will be less than 750 hours. The lamps showing less light are below standard and do not serve the purpose. Excessive gettered lamps do not have longer life as generally believed—the getters act as impurities for the filament to fail.

Effect of Voltage Variation

As discussed earlier on design aspect of tungsten filament, it is seen how more and more technicalities and sophistication are introduced for getting maximum advantages from it. A small variation in thickness, a little stretching of the filament cause a great difference in the output of light and the period of lighting. In addition if variations are made on the working pressure *i.e.*, the voltage—this would also play a very prominent role in the working of the lamp.

At present the filament is so manufactured that its optimum efficiency around 2650°C it would provided best possible spectrum when all other conditions such as production methods and ultimate power supplies are maintained within the rated standard. Whereas in India, the voltage fluctuation is so high that it can be easily termed as an abnormal condition. These semi temporary conditions are not given much importance in theoretical field but for practical purposes may seek a way out without knowing fully well about the effect of these variations.

The normal supply voltage in India is 230 volt but during peak load-period this drops to even 190 volts; on the other hand during minimum load period the voltage rises to 250 volts. The consumers often complain about the premature failure of lamps. Thinking that 250 volt filaments are stronger, some electricity boards prefer to use them though their line voltage suffers from the same fluctuating fault. Practically the 250 volt filament when burning with 190 volt supply will consume more current, give less light which may look as a mocking glow from a thin red wire serving no purpose but consuming electricity. On the other hand for a 230 volt rated filament, when the supply rises to 260 volts, it almost prove fatal to the filament as the increase and decrease are always very sudden and come without prior warning.

The effect of variations in watt intake for a gas-filled lamp is given below.

From the above curve it can be seen that a simple variation of 5% voltage the light output is affected by 20%.

A 5% increase in the voltage of a lamp, the life is reduced by nearly 30%. So, for a 230 volt filament lamp when the voltage is increased to 253 volts (which occurs many times in a day), the filament life is halved. The consumers do not know about this technical deficiency. Moreover the fluctuations of supply cause double the harm as these cause springy action in the filaments seen as 'flicks'. These flicks make the filament sag and ultimately give away from its weakest spot.

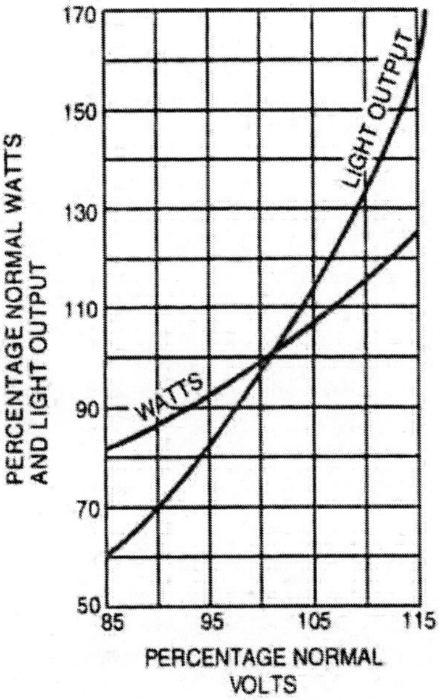

Fig. 4.12

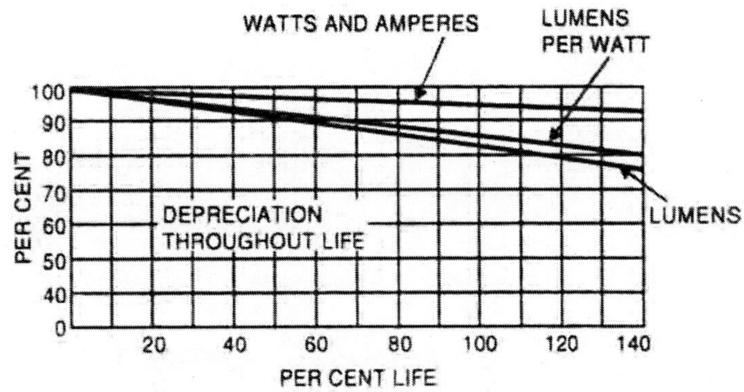

Fig. 4.13. Loss of light due to use in incandescent lamp.

The same effect of voltage variations also cause the auto headlamp filaments to fail early. In most of the running vehicles the cutout is a very neglected instrument. When the dynamo feeds the headlamps directly, the supply voltage is determined by the action of the cutout which automatically increases or decreases the dynamo field supply, the improperly adjusted cutout and the worn out dynamo brushes have always the tendency to supply sparking and flickering voltage which also light up the headlamps. In fact complete electrical system in an automobile should be in good condition to receive the best service from its lamps. The automobile owners frequently change the brand of lamps but never care to keep the few screws of the light holders properly tightened.

Due to the very high incidence of voltage variations, and at the same time without any statutory norm for the supply voltage by the electricity boards the domestic supply has been assumed by the consumers to be 220 volts to 240 volts. Strict adherence of the standards can be aimed if the supply becomes standard. It is aimed to put 240 volts as standard supply voltage in India. All the GLS lamps should be made with 240 volt filaments. The filament manufacturers should not be given licence to manufacture in the range from 200 volts to 280 volts and import of flaments other than 240 volts if stopped then a common standard can be reached.

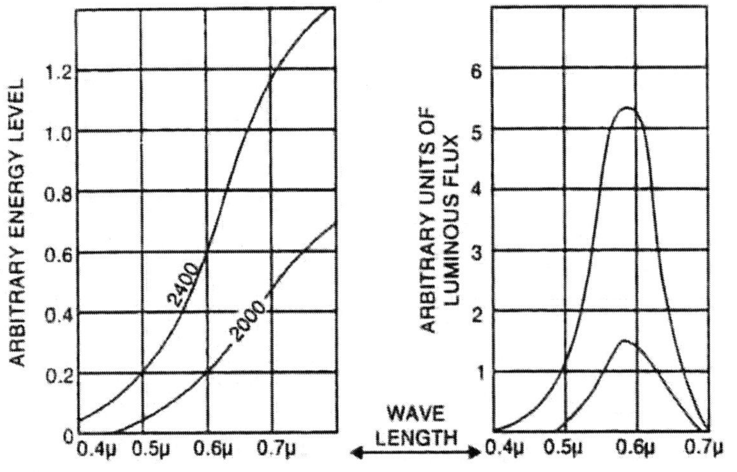

RISE OF EFFICIENCY WITH LUMINOUS FLUX

Fig. 4.14

The relation between the lamp characteristics and lamp voltage have been found as following :

Let V be the rated voltage and

v be the applied voltage, then

1. $\dfrac{\text{Lumen output}}{\text{Rated lumen output}} = \left(\dfrac{v}{V}\right)^{3.3}$

2. $\dfrac{\text{Power intake}}{\text{Rated wattage}} = \left(\dfrac{v}{V}\right)^{1.43}$

3. $\dfrac{\text{Luminous efficiency}}{\text{Rated luminous efficiency}} = \left(\dfrac{v}{V}\right)^{2.0}$

4. $\dfrac{\text{Life}}{\text{Life for normal voltage}} = \left(\dfrac{V}{V}\right)^{-13.3}$ for vacuum -14.0 gas filled

Most of the small manufacturers do not manufacture lamp components. Filaments are manufactured either by large scale manufacturers of lamps or by filament factories. Special sophisticated instruments are needed to test the characteristics and qualities of the filaments. Service of these instruments are not easily available in India. So, the filament users have a very weak and uncertain case while claiming for the best efficiency of their product. Unless the filament manufacturing is also brought under standard and control, the lamp consumers can never get satisfactory service.

Above all, the erratic voltage fluctuations both in higher and lower ranges shall always remain beyond the control of the lamp or filament manufacturers. By application of the above equations, it can be easily proved that the lamp life is mainly dependent on applied voltage for which a great improvement is necessary of our electricity production and supply system. In any case the lamp maker is always responsible for the luminous efficiency of his product and the burning life also goes with it too. So, utmost care and best of technical abilities are to be engaged by the Lamp Makers to make real good lamps against all these adversities.

REFERENCES

1. **Moon and Parry**
   The Scientific Basis of Illuminating Engineering.
   McGraw Hill Book Co., New York.
2. **Automatic Digital Photometer for Testing Incandescent Lamps**
   Philips Tech. Review Holland 23 : 244/45 : 1961/62.
3. **Principles of Illumination**—H. Cotton
   Chapman and Hall Ltd., London.
4. **Light, Photometry and Illuminating Engineering**
   William E. Barraws
   McGraw Hill Book Company Inc., New York.

5

Elementary Electrical Theories for Lamp Making

Lamp industry is a synonym to electrical industry. To put electricity for useful purpose, a lamp is made. A lamp simply is a resistance placed in the path of electric energy which starts radiating, glowing and finally giving light. Not only the functioning of the lamp, even for production of electric lamp, the main source of power is electricity. So, every lamp maker needs to be conversant at least with the basic fundamentals of electricity.

Electricity is simple term means 'the flow of electrons through a conductor'. Generally the flow of electricity through a wire is compared to the flow of water through a pipe to understand its behaviour.

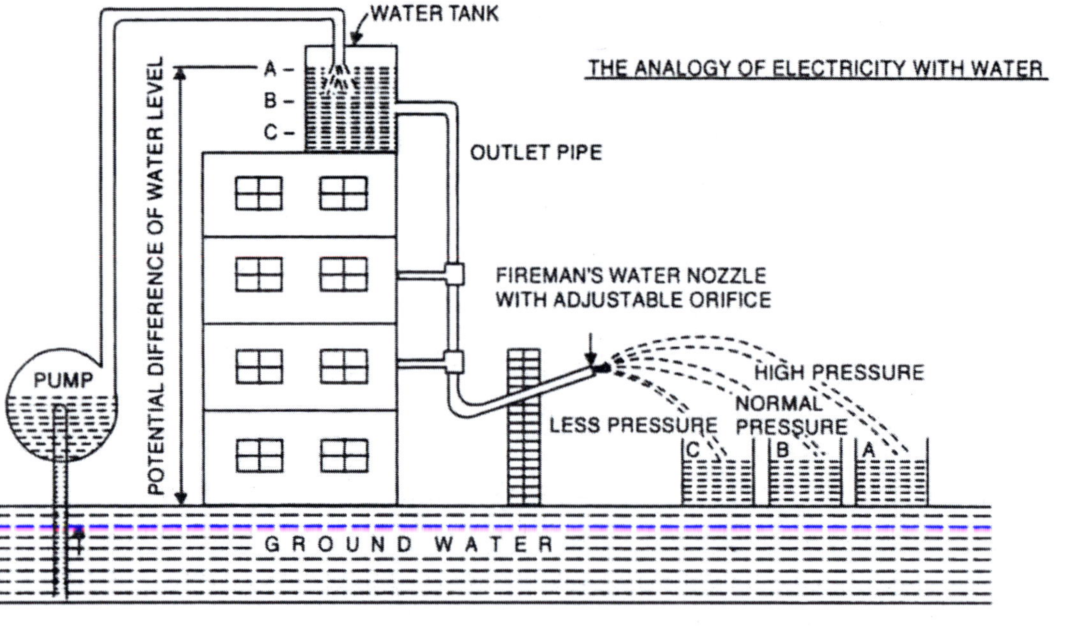

Fig. 5.1. Figure of the building.

The underground water is lifted to the overhead water tank by the pump. The height of the building causes the 'potential difference' (in short PD) between the underground water level and the overhead water tank which causes the water to flow through the tank outlet pipe under pressure. The higher the level of water in the tank, more and more would be the pressure in the pipe lines.

The Fireman's nozzle is the resistance to the flow of water and the water is thrown at a distance in the form of spray which fills the troughs A, B and C. When the level of water in the overhead tank is midway (meaning normal working level of water) the spray reaches point B; but the rising or lowering of the water level would cause fluctuations in the potential difference and the spray would then reach point A for higher level and point C for lower level. This simple analogy is known to all of us.

We now substitute the water pump with the 'generator', the pipe lines as electrical wires, name the water-pressure as 'Volt' and the quantity of water flowing out through the nozzle as 'Ampere'. We would find that electricity which cannot be seen by us, easily could be explained in this manner.

From the illustration it becomes clear how the fluctuations of pressure (volt) affects the final work of the spray going far and near. If we take the pressure at point B to be 240 volts then there is marked change in 'A' being at a lower pressure of 210 volts and at point 'C' being at a higher pressure of 270 volts. In India, voltage fluctuation is a very common phenomenon and care must be taken to design or testing of any electrical equipment of this factor. In the illustration the size of the nozzle is not changed but different results are obtained due to change of pressure in the pipe line; so, while working with photometric integrators, initial rating tests or life tests, the supply voltage is to be kept within prescribed accuracy.

The Fireman's nozzle does the useful job. The amount of water flowing through the nozzle depends on the diameter of the orifice of the nozzle. The amount of water meaning 'quantity of electrons flowing' in electrical terms is known as 'ampere'.

The nozzle is the connected load to this water circuit. Two main factors are involved for the working of this nozzle :

1. The pressure of the pipe line (Volt).
2. The quantity of water flowing per minute (Ampere).

The product of the above two gives us the rate of doing work. In electricity, the product of volt and ampere is known as 'watt'.

Now we come to another important factor of this circuit, when we know that the quantity of water flow is always governed by the size of the orifice of the nozzle. Let us take a circular adjustable type of nozzle in which the orifice can be made smaller or larger. By reducing the diameter more resistance is put on the flow of the water and this flow can be increased by enlarging the orifice. This reduction and increase of the orifice can be compared with the 'Resistance' in an electrical circuit. The thinner the wire the greater is the resistance.

In the lamp industry, we are much concerned with the resistance. The tungsten filament is a resistance wire which forms the life line of the lamp. For manufacturing and installing inside the lamp, the tungsten filament must receive utmost care and attention. A simple example of this is a coiled coil filament. The coiling of the filament is made to shorten the length but at the same time it must keep its resistance value. This is done by carefully placing of the coils which do not touch one another. A little stretching would change the resistance value and shorten the life of the lamp.

From this 'flow of water' analogy, we have now understood the following common terms in electricity :

1. Volt means electrical pressure.
2. Ampere means the amount of electrons flowing.
3. Resistance means any obstruction or resistance placed in the path of the flow of electrons.
4. Watt means the power consumed in the circuit.

PRODUCTION OF ELECTRICITY

In 1831 Michael Faraday observed that when a conductor is moved in a magnetic field, potential difference takes place in the body of the conductor and if the end of this conductor is connected externally, electrons start flowing from higher potential to the lower potential through the connecting wire which is termed as electricity.

Depending on the method of production, electricity has three types of supplies.

D.C. Supply (*Direct Current Supply*)

In a circuit where the flow of current is always in one direction it is known to have D.C. supply. The pure direct current flows as a straight line as can be seen in the following figure.

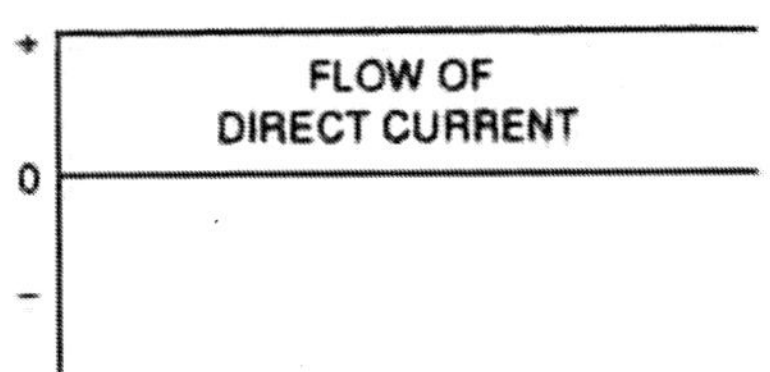

Fig. 5.2

The current does not have any cycle or frequency and it does not change its polarity.

All the lead acid batteries, torch cells produce this type of one direction flowing direct current. Electricity received from every individual battery produce a very limited voltage (pressure). So to make a powerful pressure a number of cells are connected together. For certain work specially in electronics this supply is indispensible.

For commercial supply of direct current, a rotor is driven by a external source as turbine, diesel, hydraulics etc. in a magnetic field. A schematic arrangement is shown in Fig. 5.3.

The current in the circuit always flows from B to B through the connected load in the same direction. This is achieved by using split rings and carbon brushes.

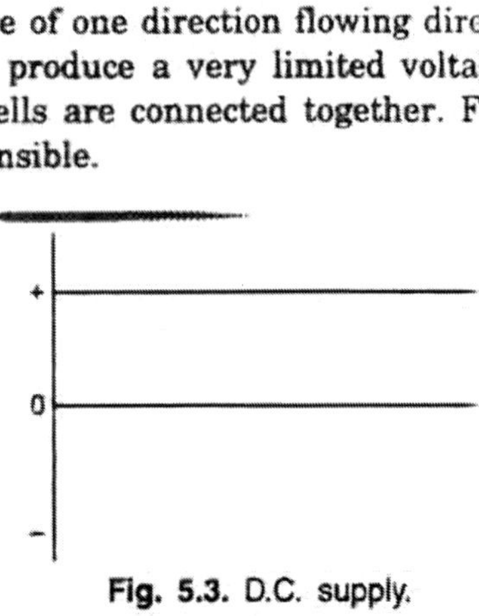

Fig. 5.3. D.C. supply.

P.D.C. Supply (*Pulsating Direct Current Supply*)

In this supply system every winding causes one impulse which diminishes to zero with further rotation of the armature. To provide a steady flow of current, a number of armature windings are introduced.

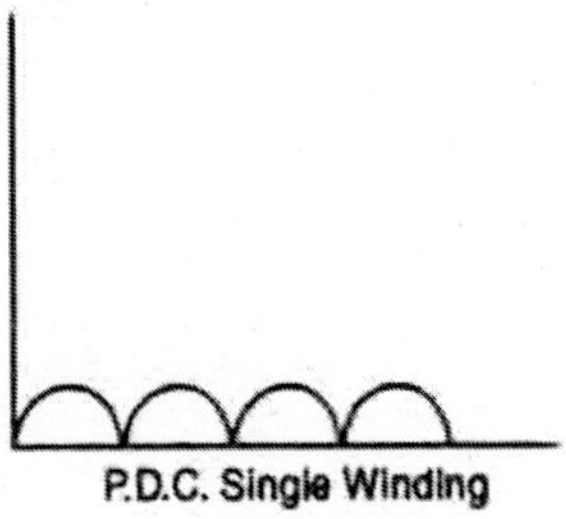

Fig. 5.4

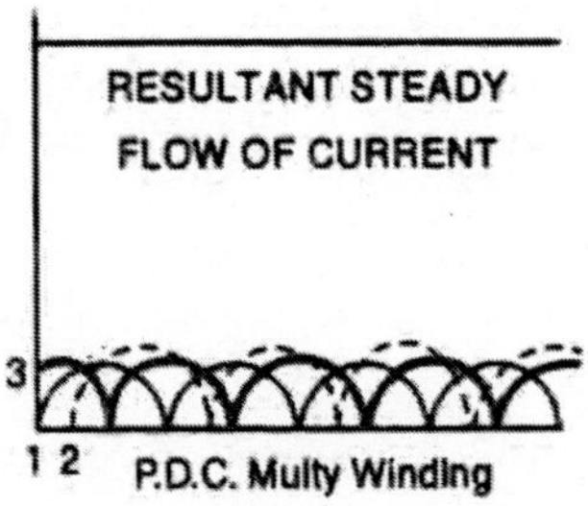

Fig. 5.5

Graphical representation of a single and multi armature windings of pulsating D.C. Winding.

A.C. Supply (*Alternating Current Supply*)

In practice it is more economical to generate electricity as alternating current.

A single turn of wire rotated in a magnetic field making one pair of poles produce electric current in the following manner. The production of current starts at zero (see in figure below), increases in one direction and reaches its peak then falls again to zero. On further rotation, the current increases on the opposite direction, reaches its maximum and returns back to zero. This set of once travel forward and reverse is known as once cycle, the current thus produced which changes it's polarity and magnitude in every cycle is known as alternating current.

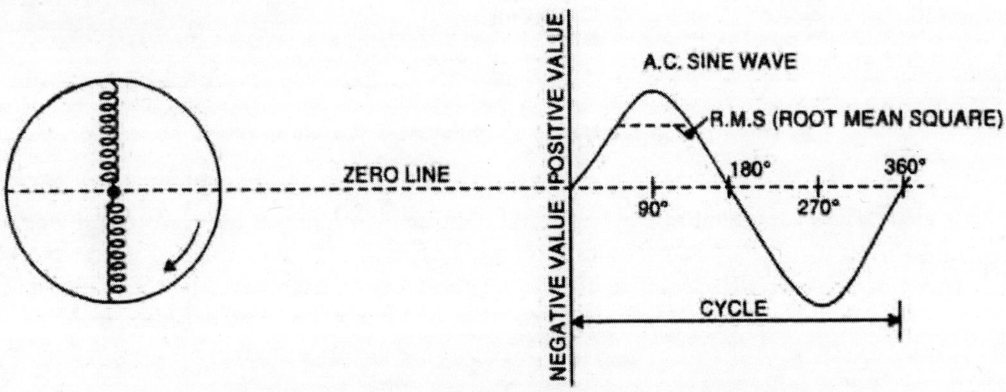

Fig. 5.6. A single phase alternator and its wave form.

The cycle is known as the frequency of the voltage. The frequency depends on number of pair of poles and the speed of rotation. The accepted frequency of our country is 50 cycles per second. If the frequency drops below 50 cycles per second then the lamp will not show a continuous light—it will start flickering.

Three Phase Supply

A three phase supply means, three separate single phase supplies are taken from three pairs of conductors of one alternator. In this system the windings in the conductor are given a 'star connection' as can be seen in the figure below.

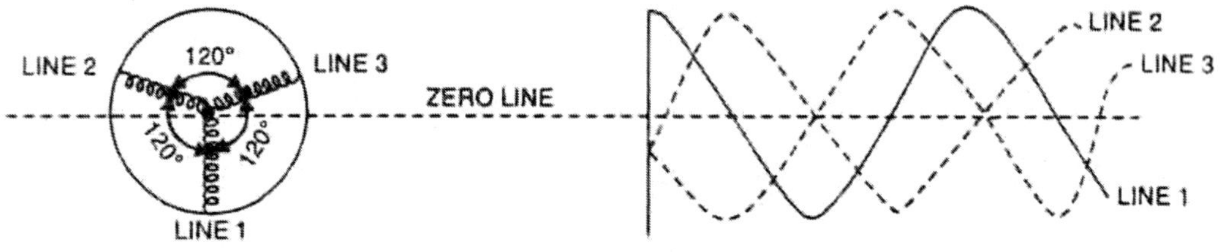

Fig. 5.7. A three phase alternator with its wave forms.

The wirings are placed 120° apart. Every wiring produces 240 volts. The voltage between the neutral and the line conductor is 240 volts and is known as 'phase voltage'. The voltage between any two line conductor is 1.73 times the phase voltage which equals to 415 volts. In the factories with 3 Phase Supply, the connection is given in the following manner.

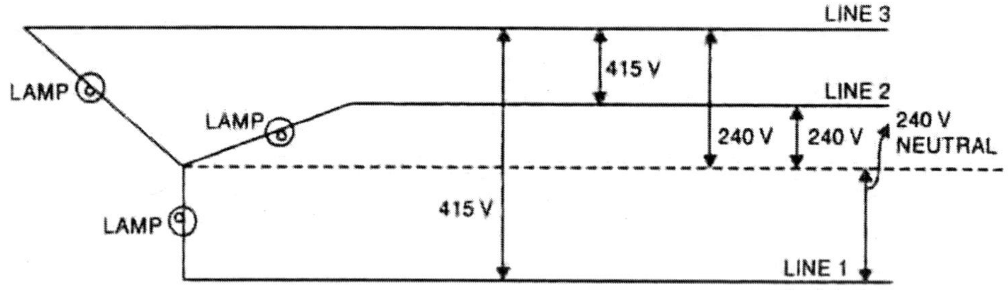

Fig. 5.8. Three phase supply giving both single and three phase service.

All the lights, single phase motors and testing equipments are connected between a phase and the neutral. At times due to wrong connections the neutral line might get supply which must be avoided. The load in the factory is to be distributed in such a manner that these are equally balanced on every phase; this is known as 'Phase Balancing'.

POWER FACTOR

Earlier it is stated that the work done by electricity is measured in 'watt' which is obtained by multiplying voltage and ampereage in the circuit. As soon as a switch is put 'on' the electrical circuit comes to life instantly; but this is only an apparent affair. In actual work; the current always lags behind voltage due to many factors as faulty motors, sparking, carbon brushes, bad insulations, unbalanced phase connections, faulty wiring etc. These deficiencies do not permit the current to flow in phase with the volt. So, the actual power consumed in the circuit will be always different than the normally calculated load. This can be clearly seen in the following graphical representation of the power factor.

$$\text{Power factor} = \frac{\text{True watt}}{\text{Apparent watt}} = \cos \theta$$

From the figure the apparent angle of lag between volt and ampere can be determined. Power factor is calculated from the cosine of the angle of lag (the cosine value can be found in the Trigonometrical Table). It also should be understood that the power factor can never be more than unity. In most ideal condition it can be 0.9% to 0.95%.

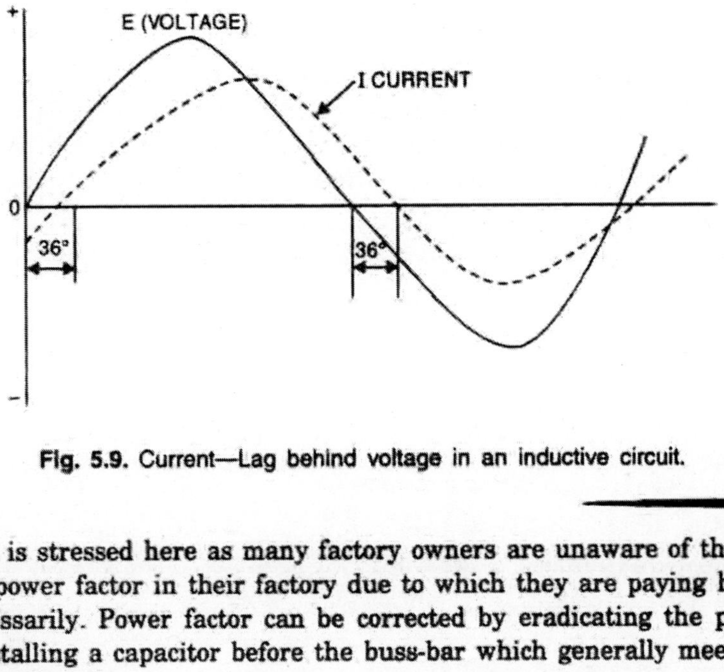

Fig. 5.9. Current—Lag behind voltage in an inductive circuit.

This point is stressed here as many factory owners are unaware of the fact that they have very low power factor in their factory due to which they are paying heavy electricity charges unnecessarily. Power factor can be corrected by eradicating the prevailing faults and also by installing a capacitor before the buss-bar which generally meets the demands of the Electricity Boards.

The Advantages and Disadvantages of the Supply Systems

1. A.C. Supply brought new dimensions to electricity when it could be carried over long distances as the line voltage loss is negligible.

2. With 3-phase system more power can now be obtained though the cost of single phase A.C. or D.C. motor is almost same.

3. Conversion of A.C. to D.C. is easier and cheaper.

4. To reduce current in D.C. system, a resistance is to be used incurring heavy losses of energy whereas in A.C. a simple choke or capacitor placed in the circuit does the same job without loss of any appreciable energy.

5. For conducting A.C. several strands of wires are to be used which entails more cost as A.C. conducts on the skin of the wires whereas in D.C. a single wire is sufficient.

6. In electrolytic process A.C. supply cannot be used. Large chemical factories where different chemicals are separated by electrolysis, all types of electroplating, electrotyping and electrodeposit jobs are carried out, only D.C. supply can be used.

7. For electronics pure D.C. supply can be used in most cases.

So, both the system of supplies have their own advantages and importances.

PRACTICAL ELECTRICITY

(With reference to lamp making industry)

Ohm's Law of Resistance

Electrical flow is governed by three fundamental factors :

1. The pressure 2. The current 3. The resistance.

In 1827, the German mathematician, George Simon Ohm first enunciated the formula of relationship between these three characteristics of electricity. He found that the 'electromotive force' (E in short) of a circuit always equals the product of the Intensity of the current (I in short) with the resistance (R in short) present in the circuit.

The Formula is $E = I \times R$

or Volt = Ampere × Resistance

So we have $E = I \times R. \quad I = \frac{E}{R}. \quad R = \frac{E}{I}.$

(Remember East Indian Railways—*EIR*—where E equals I into R).

Units of ε E.I.R.

E : The unit of electromotive force (e.m.f.)

ε or potential difference p.d. is *Volt.*

Volt. The potential difference between two points is 1 volt if 1 joule of work is done in moving 1 coulomb of charge from one point to other.

$$1 \text{ volt} = \frac{1}{300} \text{ e.s.u.} = 10^8 \text{ e.m.u.}$$

1 volt is equal to $\frac{1}{1.083}$ e.m.f. on a Weston cadmium cell at 20°C.

I. (Intensity of current). The unit of current is ampere.

Ampere. The current flowing through a conductor is one ampere if the rate of flow of charge is 1 coulomb *i.e.,* 3×10^9 e.s.u. of charge per second.

In International Standard, it is that steady current which flowing through a voltameter in one second would deposit 0.001118 gm silver.

R : (Resistance). The unit of resistance is ohm. Symbol Ω.

Resistance. A conductor has 1 Ohm resistance if a current of 1 Ampere flows through it under potential difference of 1 volt.

One International Volt is again equal to the resistance of a column of mercury 106.3 cm. long and 1 sq. mm. Cross-section at 0°C (when mass of mercury is 14.452).

When highly accurate readings by Photometric Integrators are taken the following conversions are used :

1 International Ampere = 0.999835 Amp

1 International Ohm = 1.000495 Ohm

1 International Volt = 1.00033 Volt.

Resistances in Series. When two or more resistances are connected end to end—one after the other as shown in the figure below, these are known to be in 'series connection'.

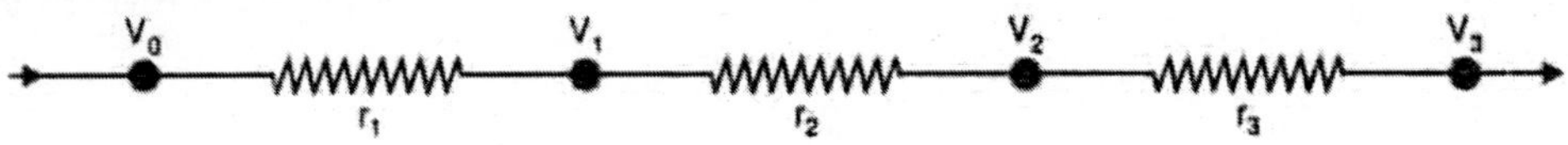

Fig. 5.10

In this type of connection, the total resistance in the circuit is the sum total value of all the resistances added together but the mains voltage V_0 will vary after every resistance (r_1, r_2, r_3) at V_1, V_2, V_3. The voltage will vary as per the Ohm's law of resistances. $V = I \times R$.

Resistance in Parallel. When one end of all the resistances are bunched and joined together and similarly the other ends also are bunched and joined together as shown in the figure below, the connection is said to be in parallel.

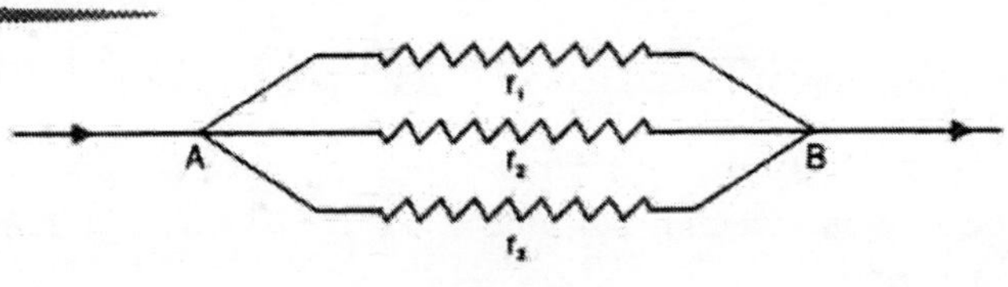

Fig. 5.11

In this type of connection the supply voltage at points A and B will not vary. In parallel connection the voltage does not vary. The parallel connections follow the rule that 'the reciprocal of the actual resistance in the circuit (known as conductance) is equal to the sum of the reciprocals of the individual resistances. In the above figure the current will be divided into three parts from point A as per the demand of the resistances which would still follow Ohm's Law. The formula for series connection is as following :

$$R = \frac{1}{\text{(Total resistances)}} = \frac{1}{r_1} + \frac{1}{r_2} + \frac{1}{r_3}$$

For example, in the above circuit if we take $r_1 = 20$ ohm, $r_2 = 30$ ohm and $r_3 = 50$ ohm, then

$$\frac{1}{R} = \frac{1}{20} + \frac{1}{30} + \frac{1}{50} = \frac{31}{300}$$

$$R = \frac{300}{31} = 9.7 \text{ ohms (approx.).}$$

Example 1. *A car uses a 6 volt 3 watt tail lamp. What is the current consumed by the lamp and what is the resistance ?*

Sol. Watt = Volt × Current   Volt = Current × Resistance

$3 = 6 \times ?$ $6 = 0.5 \times ?$

or $\frac{3}{6}$ = Current   or $\frac{6}{0.5}$ = Resistance

or = 0.5 amp. current   or = 12 ohm resistance

The tail lamp has 12 ohm resistance and consumes 0.5 amp. current.

Example 2. *Find out the resistance of a 12 volt 36 watt auto head lamp.*

Sol. Formula. Watt = Volt × Ampere   Volt = Ampere × Resistance

$36 = 12 \times ?$ $12 = 3 \times ?$

$\frac{36}{12}$ = ampere   or $\frac{12}{3}$ = ohm.

= 3 amp.   or = 4 ohms

The head lamp consumes 3 amps. current and has 4 ohms resistance.

Example 3. *What is the burning filament resistance of a 250 volt 100 watt GLS lamp ?*

Sol. Formula. Watt = Volt × Current   Volt = Amp. × Resistance

$100 = 250 \times ?$ $250 = 0.4 \times ?$

or $\frac{100}{250}$ = 0.4 amp. $\frac{250}{0.4}$ = 625 ohm resistance.

The GLS lamp consumes 0.4 amp. current and has a burning resistance of 625 ohms.

Example 4. *With M 6/2 type miniature lamps a chain (series connection) is made of 48 lamps. What is the current consumption of the chain ? Justify the numbers to be connected to a 250 volts mains supply.*

Sol. There are 6 volt mini lamps consuming 1.5 watt power

1.5 watt = 6 volt × Current

or $\frac{1.5}{6}$ = 0.25 amp. current.

Because the lamps are connected in series the current will remain unchanged in the series. But the filament of the mini lamp can take only 0.25 amp current, so care should be taken to see that the current intake in the chain should never increase. From Ohm's law we know that voltage is directly proportional to the current which means any increase in voltage supply will increase the current in the circuit.

When 48 of these mini lamps are connected in series, these would withstand the electrical pressure upto 48 × 6 = 288 volts. If the line voltage is 240 volts then 41 lamps would be the ideal number of lamps. The extra 7 lamps are added to suffice for 15% voltage fluctuations on higher side so that the lamps will not fuse even when the line voltage is approximately 290 volts. Of course more than 50% light value has to be sacrificed due to this safety measure. It would be better if this series is made with 45 lamps which would withstand 10% voltage fluctuations and provide more light.

Total power consumption :

As the current does not vary in a series connection. The total power consumption will be 250 volts × 0.25 amp = 62.50 watt.

Example 5. *If 48 nos. 100 watt/250 volts GLS lamps connected in parallel are to be tested in an Ageing Rack, how much current will be flowing through the circuit when 10% over voltage will be applied ? What capacity variac would you recommend for this testing ?*

Sol. $W = V \times I$ $\qquad$ $E = I \times R$

or $100\text{ W} = 250\text{ volt} \times I$ $\qquad$ or $250\text{ volt} = 0.4 \times R$

or $\dfrac{100}{250} = I$ $\qquad$ or $\dfrac{250}{0.4} = R$

or $I = 0.4\text{ amp.}$ $\qquad$ or $= 625\text{ ohm resistance}$

Reciprocal of conductance (total resistance) = $\dfrac{1}{R}$

Each individual resistance equals the sum of $\dfrac{1}{r_1} + \dfrac{1}{r_2} + \dfrac{1}{r_3} + \ldots\ldots + \dfrac{1}{r_n} = 48$ (lamps)

So for the GLS lamps $\dfrac{1}{R} = \dfrac{1}{625} \times 48$ ohm = 13.02 ohm.

Total resistance in the circuit is 13.02 ohm.

Calculations for total current flow at 10% over voltage.

Line voltage 250 volts. 10% of 250 volt is 25 volts.

Total volt 250 + 25 = 275 volts

Formula. $V = I \times R$

or $275\text{ volt} = I \times 13.02$

$\dfrac{275}{13.02} = 21.12$ amp. current = Total load on the circuit.

So, the variac must have the capacity to take atleast a minimum load of 25 ampere.

Example 6. *Two lamps each having 120 ohms resistance are connected in parallel, in series with them is another lamp of 40 ohms resistance. All these three lamps are connected to a supply served by 240 volts. What is the total resistance in the circuit ? What is the current consumed by each lamp and what is the total current in the circuit ?*

Sol. Part I

The conductance of the two lamps in parallel

$\dfrac{1}{R} = \dfrac{1}{r_1} + \dfrac{1}{r_2}$ or $\dfrac{1}{R} = \dfrac{1}{120} + \dfrac{1}{120} = \dfrac{2}{120} = \dfrac{1}{60}$, So $R = 60$ ohms.

Part II

The total resistance in the circuit is 60 ohm + 40 ohm = 100 ohms

Formula. $I = \dfrac{E}{R}$ when $E = 240$ volts, $R = 100$ ohms

$$I = \frac{240}{100} = 2.4\text{ ampere}$$

So, total current flowing in the circuit is 2.4 ampere.

Further Laws of Resistance *(as applied to lamp industry)*

1. The resistance of a filament wire is directly proportional to its length.

2. The resistance of a filament wire is inversely proportional to its cross-sectional area.

3. The resistance of the filament is taken as the resistivity of tungsten (It is 5.51×10^{-6} ohm cm cube at 20°C).

4. The resistance of the filament will depend at the temperature it is working.

The first three laws being physical in nature, are calculated as per the following formula :

$$R = \rho \times \frac{l}{A}$$

where R = resistance of filament

(Rho) ρ = resistivity of tungsten

l = length of the filament

A = area of cross-section of filament.

The fourth law dealing with temperature actually is the temp. co-efficient of a conductor (here tungsten) which is a constant and is denoted by α (alpha).

Example 7. *If the resistance of a cold filament is 250 ohm and its length is 750 mm. What should be the diameter of the filament ? (when tungsten resistivity is 5.51 micro ohm per cm cube at 20°C cold condition).*

Sol. Formula. $R = \rho \times \frac{l}{A}$ so $A = \rho \times \frac{l}{R}$

We have $R = 250$ ohm

$l = 750 \text{ mm} = 75 \text{ cm.}$

$$= \frac{5.51}{1,000,000}$$

$$\text{(Alpha) } \alpha = \frac{\overset{3}{\cancel{75}} \times .00000551}{\underset{10}{\cancel{250}}} = \frac{.00001653}{10} = \frac{.00000551 \text{ ohm}}{.000001653 \text{ sq. cm.}}$$

So the diameter of the filament will be

The area of the cross-section of the filament is $\frac{\pi d^2}{4}$

So we have $\pi d^2 = .000001653 \text{ sq. cm} \times 4$

$$d^2 = \frac{.000001653 \times \cancel{4}}{\underset{.785}{3.\cancel{141}}} \qquad \text{(when } \pi = 3.141)$$

$$d^2 = .000021 \text{ sq. cm}, \quad d = \sqrt{.000021}$$

or $d = .004 \text{ cm} = .004 \times 10 \text{ mm} = .04 \text{ mm}.$

Here from practical measurements, it has been found that in a 230 volts, 100 watt lamp the filament length is about 1 metre and its diameter is about .05 mm, the diameter of the coil is 0.3 mm. The formula of resistance given above is very important to the lamp manufacturers when they have to locally make filaments of different voltage and wattage by adjusting only the length of the filaments available with them.

USEFUL FORMULAE FOR LAMP INDUSTRY

(*a*) $E = I \times R$, $I = \frac{E}{R}$, $R = \frac{E}{I}$

(*b*) $W = E \times I$

(*c*) $W = (I \times R) \times I$, So $W = I^2R$

(*d*) $W = E \times I$, $W = E \times \frac{E}{R}, W = \frac{E^2}{R}$

(*e*) In series connection $R = r_1 + r_2 + ... + r_n$

(*f*) In parallel connection $\frac{1}{R} = \frac{1}{r_1} + \frac{1}{r_2} + ... + \frac{1}{r_n}$.

For all above formulae E = Volt (Electromotive Force or Electrical Pressure)

I = Ampere (Intensity of Current)

R = Ohm (Resistance)

W = Watt (Product of Volt and Ampere)

(*g*) Formula of filament resistance calculation

$$R = \rho \times \frac{l}{A}$$

where R = resistance of the conductor

ρ = resistivity of the conductor

l = length of the conductor

(alpha) α = area of cross-section of the conductor.

6
RAW MATERIALS FOR LAMP MAKING

In lamp making the materials used mainly are :

A. Metals B. Sealing wires C. Glass.

A. METALS

1. Tungsten
2. Molybdenum
3. Nickel
4. Aluminium
5. Solder.

1. **Tungsten.** Of all the metals used in an incandescent lamp, tungsten is the most important metal which takes up the hardest working load. It is tungsten which provides the necessary resistance to the flow of the electric current and at the same time retains its own physical structure without getting burnt out. So let us see how and why tungsten came to be the precious child of the lamp makers. First of all let us study the following curves of spectral power distribution of a metallic body at different temperatures in a black body radiator.

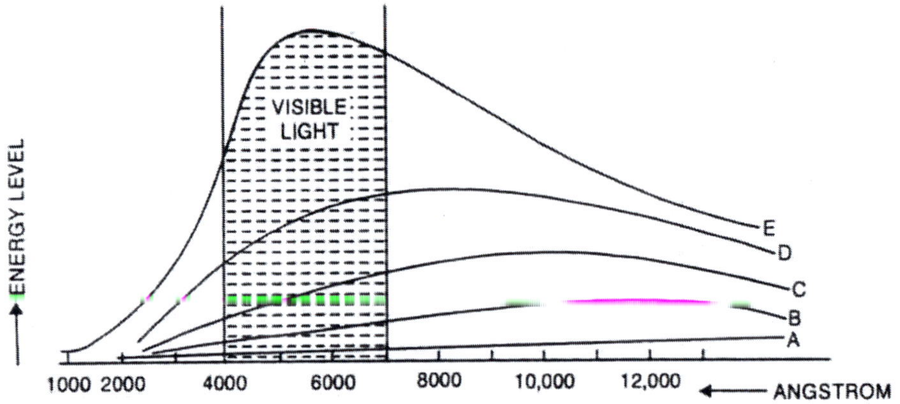

Fig. 6.1

In the curve A for 2000 K, the visible light energy between 4000 Å to 7000 Å is very less. Most of the energy in this curve is spent on producing heat in Infrared region and whatever little visible light it has made, it is mostly red in colour.

Whereas in curve E, there is sudden rise of energy level in the visible light region and also sudden drop in the ultraviolet region.

So, we have to search a metal for the filament which could produce visible energy if not like curve E at least like B, C or D.

From the above curves it is also seen that with the increase in temperatuse how the predominant red colour in lower temperature changes to day-light which is the optimum necessity of our eye. So, while selecting a material for the filament, we shall have to consider two things.

(*a*) It should be able to withstand greatest possible heat so that it could produce more visible light approaching to day light. In terms of energy consumption, this must be able to produce more light per watt of energy consumed—so that it will make cheaper light.

(*b*) The colour of light also should not be distinguishable—meaning it should not favour any colour spectrum but it should be as white as the day light.

From chronology of 'lamp making', we have seen, it was mostly the selection of the materials for the filament which made improved lighting. From the above black body radiator curves, it is evident that any material having melting point below 2000°C would not suit the lamp makers. But during the early days of lamp making, these datas were not available with the pioneers of lamp making.

Iron and steel melt at temperature between 1300°C to 1600°C, C still the Westinghouse 'stopper' filament created wonders in 1893 A.D. and gem or metallised carbon filaments made a great impact on the first users of electric lamps.

The following materials have been used as filament from time to time to produce electric lamps commercially.

	Melting point
Platinum	1773°C
Osmium	2700°C
Tantalum	2900°C
Tungsten	3380°C
Carbon	3600°C

Till today, the main problem of the filament is slow evaporation at its working temperature which is much lower than its melting point. With the increase in temperature the rate of evaporation also increases. Of all the above materials carbon has the highest melting point but it cannot be worked at temperature more than half it's melting point. At 1850°C, the carbon filament gives only 2/3 lumen per watt and the wall of the lamp gets completely blackened even before the filament gives away.

Platinum was the first metal to be tried as filament but had to be dispensed with due to very high rate of evaporation and low lumen output. In 1901, Siemens Ltd., used tantalum in high voltage lamps. But the disadvantages of evaporation causing low lumen output per watt continued and the last metal tungsten replaced all other materials. It is doubted if any other metal will be able to replace tungsten as filament material for electric lamps.

The alphabetical letter symbol of tungsten 'W' is taken from the name of its ore as Wolframite. Tungsten also is available in ore form as Scheelite. Through many stages of chemical processes, all the impurities in the Wolframite ores are eliminated to get chemically pure 'tungsten trioxide'. Just and Hanaman in U.S.A. first substituted tungsten for carbon. In the preparation of tungsten for filament making, very high degree of purity is essential as each and every portion of the filament is put under hardest duress to produce light. It was Coolidge between 1906 A.D. to 1909 A.D. who started experimenting with tungsten purification and succeeded in making the filament. It has been seen that after switching on the lamps, within fraction of a second, due to excessive heat the tungsten molecules grow into bunches of crystals. While this crystal growth takes place, the filament wire lengthens causing the filament to sag. In the early stages it took several years to detect the cause of this sagging and the lamps failed due to this cause. After discovering this crystalline effect, two compounds of Tungsten were formulated to overcome this difficulty.

1. Oxides of thorium, uranium and aluminium which act as inhibitors of grain growth on individual particles of tungsten powder were added during the sintering process. This types of compound prevents crystalline growth taking place and the filament has a fine grain structure. This was the common practice for a long time in filament making. Till today filaments for rough service vacuum lamps, this process is being adopted.

2. The main reason for failure of a sagged filament was due to the presence of molecules of impurities which got sandwitched between the boundaries of the tungsten crystals. Modern wire for coiled coil filament of general service lamp is made from tungsten powder in which small quantities of aluminia or silica is added. These additives have profound effect on the structure of the resulting tungsten wire by assisting in the clean up of the crystal boundaries which promote rapid grain growth both during the sintering process and in the finished tungsten filament.

The chemically pure tungsten trioxide is added with either of the above two additives and heated in a tube furnace in hydrogen atmosphere at about 1000°C. The wolframite, now as pure tungsten trioxide, becomes pentaoxide, to dioxide and reduced finally to pure Tungsten in the form of a sooty grey powder as

$$WO_3 \text{ (reduced to)} \rightarrow WO_2 \text{ (finally reduced to)} \rightarrow W$$

The sooty grey powder of tungsten is then pressed inside square dies of 0.65 sq. cm cross-section and 40 cm in length. This bar is then sintered by passing through it an electric current at the high current density of 4000 amp per sq cm. During this sintering process, the bar is again surrounded by an atmosphere of hydrogen to prevent any oxidation. Before sintering, the tungsten was a very fragile bar made from loose tungsten soot powder, becomes now very hard as glass brittle metallic rod. The following would give some indication on the changes brought in the tungsten through sintering.

Physical state	*Length cm*	*Width cm*	*Depth cm*	*Mass gm*	*Density gm/cc*	*Appearance*
Before :						
Sooty tungsten						
Powder	40.0	0.8	0.8	290	11.20	Dark grey
After sintering :						
Glass brittle, hard	35.0	0.7	0.7	285	17.36	Bright silver

The sintered bar being ductile is hammered into thin rod in a series of swaging machines (*See Photo 6.1*) till the diameter is reduced to 1.8 mm and the length nearly 5 metres. The swaged wire then is drawn in a chain drawing bench through several operations for removing the surface irregularities. These preliminary drawings for removing surface irregularities are of great importance as these irregularities if not eradicated become the centres for crack formation. As more and more operations are carried out, the tungsten becomes more and more ductile. One of such rod can produce several miles of filament wire while drawn through 'diamond dies' even upto a diameter of 10 microns. But at such a thin state also, it has a tensile strength several times more than that of steel. At the final stage, the tungsten becomes so ductile that it can be wound round a mandrel even upto 3/4 time of its diameter.

For shortening the filament length, the filament is wound round a mandrel. By this process one metre of Tungsten filament is reduced to a length of around 2.5 cm. In case of single coil filaments the mandrel is a steel wire and in the case of coiled coil filament this is made of molybdenum. During heat treatment process, these mandrels are dissolved away leaving the coiled filament. Heat treatment of the filaments are also carried out in hydrogen atmosphere for annealing at 800°C to 900°C for single coil filaments and 1600°C to 1700°C for coiled coil filaments.

Previously single coiled filaments were used in gas filled lamps. Attempts were made to reduce this size further so that the heat could be more concentrated to a smaller area and the loss of heat through convection would be reduced. In 1930 Benbow made the first coiled coil filament. This is achieved by first coiling and reducing the length of a coil and later coiling this single coiled filament into a spiral. This reduced the 50 cm length of a coiled filament to approximately 20 cm length. Where 5 to 7 filament supports were necessary; now only two supports were able to carry the filament avoiding loss of heat by 5 supports and also reduce their cost. The first non-şag Tungsten filament was made by Dr. Aldar Pacz of U.S.A. after 218 trials.

Coiled coil filaments give 20% greater efficiency over single coiled filament. With the decrease of the size of the filament wire, the wattage of the lamp is increased. For a 100 watt lamp, the filament diameter is approximately 40 Microns and length 900 cm which is coil coiled to the size of about 25 cm.

(*See Photo 6.2*)

Design Factors of Tungsten Filaments. The cross-section of the filament determines the current rating of a lamp and it varies as the square of its diameter which radiates energy and at the same time comes in contact with the inert gas. This is the reason why a simple extra stretching of the filament allows excessive current to flow and the filament fails prematuredly. In India, the Standard Line-voltage is 240 volts (presently varying between 220 V to 250 V) and all our machineries and wirings are standardised and there is very less chance of changing them. But, while designing filament it is to be remembered that with the decrease of the voltage for the same wattage lamp, the filament must be made larger to take up the increased current and in such an event only the lamp will become more efficient. Regardless of the size and shape of the filament, the filament is always surrounded by a stationery sheath (Langmuir sheath—it is described later) of hot gas nearly of constant thickness. Therefore heat losses from the filament is less as the length of the filament is increased and filament evaporation is reduced. So, longer filament can operate at higher temperature at lower voltage with higher efficiency for the same rated life.

Tungsten has a positive temperature co-efficient of resistivity. In cold condition the resistance of the Tungsten filament is 1/15th to 1/17th lower than the final operating resistance. So, whenever a light is switched 'on' there is 'on rush' of current. It should be remembered that the filament dissipates its energy by conduction through surrounding gas, electrodes, support wires and the bulb base. With the increase of the filament temperature more amount of visible light is produced and this temperature also provides more efficiency on wattage as the heat losses through convection and conduction also diminishes. For a long time, lamps below 40 watt were made vacuum lamps. With the increase in the efficiency of tungsten filament, even 25 watt gas filled coil of coiled filaments are becoming regular in service.

The Tungsten filament is the 'life' of the lamp. Almost all the studies in lamp making are concentrated and directed towards improvement of the filament only. Tungsten wire can be drawn to a diameter of 10 micron (.0004 inch) which is about one sixth diameter of a human hair. A definite specification of filament diameter, its length, form, coil spacing, number of filament supports, method of mounting all of which are related to filament are to be taken into consideration when designing a lamp. The other items related with the filament are lead-in-wires, gas, gas pressure, bulb size, shape and temperature. From these physical factors arise the final objective of lamp design to provide accurate wattage, to have uniformity and standards, efficiency and life rating which all combined together will ensure dependable service to the user. The following gives an operating performance of a general lighting service lamp.

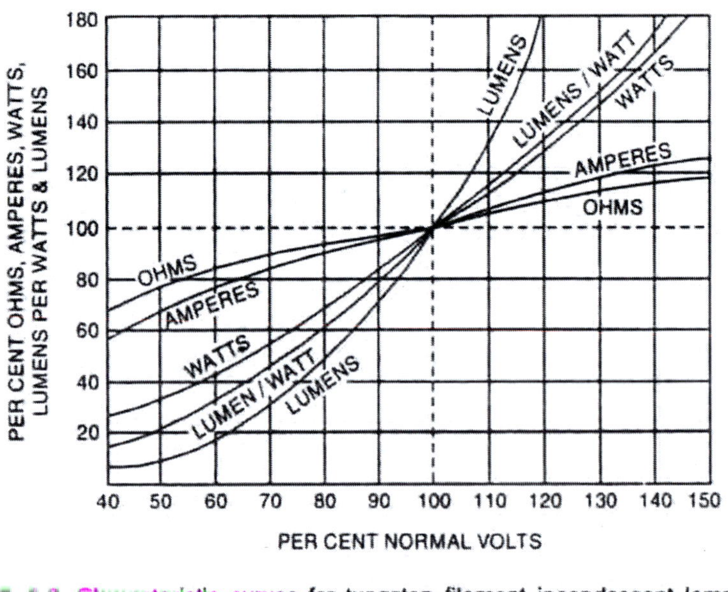

Fig. 6.2. Characteristic curves for tungsten filament incandescent lamps.

For testing all lamps, the lamp is burnt with base up in a vertical position. The base temperature is measured at the metal base covering the neck or the seat of the lamp. The ambient temperature for taking the measurements in a lamp is 25°C. From the above curve, it is seen how a constant potential (meaning the voltage supplied to a lamp reasonably remains constant) filament looses its light gradually. As the lamp burns, the filament evaporates which reduces the light and at the same time the cross-section of the filament gets reduced due to evaporation. The filament gets thinner and gives more resistance to the flow of current ultimately producing less light.

2. **Molybdenum.** Molybdenum is an extensively used metal in the lamp industry. With its high melting point of 2610°C, it exhibits high strength at elevated temperature. It is used for filament supports and mandrel wire. Molybdenum has co-efficient of expansion that matches the co-efficient of expansion of some categories of glass and vitreous materials; so it is used at many places as glass-metal-seal as seen in halogen lamps. Molybdenum is best suited for support wires and hooks as it has lower ductile-brittle transition temperature than tungsten.

3. **Nickel.** Nickel is a scarce metal in India. It has very extensive use in the industry. In the lamp industry it is used for electroplating iron and steel. A very small percentage of nickel with iron (steel) gives it great hardness and tenacity. The nickel steel combination is highly resistant to oxidation even at red hot condition. So, nickel is used in the lead-in-wires, as it avoids scaling in heated condition. This is also used for bi-metal strips in the starters for fluoroscent tubes.

4. **Aluminium.** In electrical industry, aluminium is slowly replacing copper. Next to oxygen and silicon, aluminium occurs most abundantly in nature. Aluminium is a good conductor of electricity.

Previously all lamp caps were made of copper. Now due to its highly malleable and ductile qualities, easy availability and cheapness in price, aluminium caps are used in lamps where ever possible. The main advantage of aluminium cap is that it forms a thin surface of oxide and it does not corrode in normal weather condition. Aluminium cannot resist corrosion under salty weather and research work is in progress to overcome this difficulty.

5. **Solder.** It is an alloy of tin and lead. It has the quality to hold on to a metallic surface. Where liquid 'killed spirit' is used as flux for soldering the lamp-cap eyelets, these tend to flow on the sides of the cap. This flux is highly corrosive and this must be cleaned with a moist cloth and dried clean immediately after soldering. As far as possible, soldering with 'killed spirit' is to be discouraged. Flux impregnated solder should be used for soldering of eyelets. Antimony is used in solder to retard the transformation of tin into its gray state, called as 'tin paste'. Normally 50 B grade solder with antimony is good quality solder for lamp-eyelets.

B. SEALING WIRES

This is one of the major component in the lamp industry. It is generally called lead-in-wire or electrode. As indicated by its name, this leads current into the lamp or this is the only media of contact between outside atmosphere and the hermetically sealed inner parts of the lamp.

Firstly it must have nearly, the same co-efficient of expansion of the glass through which this passes. For soda glass, platinum would be most suitable metal but its high cost and brittleness tendency at high temperature renders it unsuitable as lead-in-wire for lamp. The most typical part of the lead-in-wire is the portion inside the glass pinch known as 'Dumet'.

Dumet. It consists of a nickel steel wire made from crucible casting ingots of 42% to 58% nickel steel. The special ingot is heat treated in hydrogen atmosphere and the alloy prepared for cold drawing. This very thinly drawn wire serves as a core over which a seamless copper tubing is enclosed. To ensure a compact joint between the core and the copper sleeving, thin brass sheet is embedded inside which acts as a flux or bonding material. After this the rod is swaged, brazed and drawn through fine dies of required size and treated with Sodium Borate.

For making a leak-proof hermetic seal at the pinch, the glass must "wet" the wire. The sodium borate prevents formation of spongy cuprous oxide while melting in the stem head, and at the same time this oxide dissolves with the molten glass. This mixture of sodium borate with glass makes a homogeneous material and covers the copper wire all around. As described earlier, because of different types of materials (as alloy, steel, copper, brass) having different co-efficient of expansion the thermal expansion of the dumet is different in radial and axial directions.

In the radial direction, the expansion of the dumet matches that of the glass and in the axial direction it is less than the glass. For this reason only the stem is not fully annealed but cooled in steps and brought down to room temperature quickly to retain a certain percentage of the axial stress. For this reason, also the dumet cannot be made of more than 0.8 mm diameter.

A lead-in-wire is composed of three parts. All the three parts—the outer lead, sealing dumet, the inner lead, are spot-welded in percussive condenser discharge type welding machine.

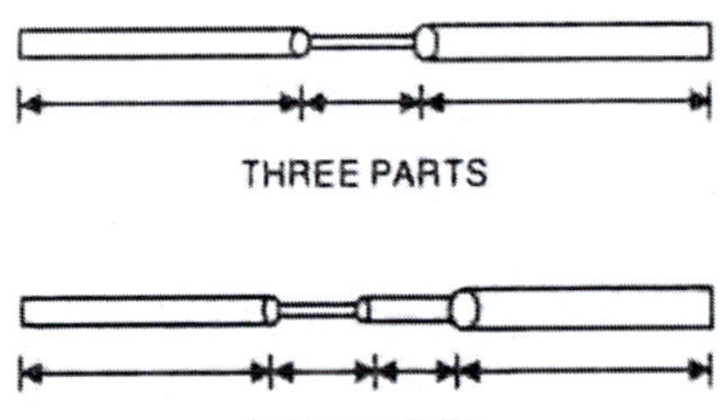

Fig. 6.3

The Outer Lead. This is generally made of oxygen-free copper for low wattage lamps. In high wattage lamps this is made of monel metal (70% Ni and 30% copper) with diameter from 0.125 to 0.50 mm. This is generally termed as 'fuse wire'. In every high watt lamp (from 60 watt onward) at least one of the lead-in-wire should be of fuse type. This is used as a safety precaution. Monel melts out and disconnects supply when a faulty lamp might overheat the base endangering electrical fire.

The Sealing Dumet. These are borated copper clad dumet wires in diameter from 0.20 mm to 0.80 metre. For 100 watt lamps it is advisable to use lead-in-wires with glass sleeved dumets.

The Inner Lead. The brunt of the whole process in lighting up the lamp is borne by this inner lead. For GLS lamps nickel plated iron wire (iron armco core) with 15% weight

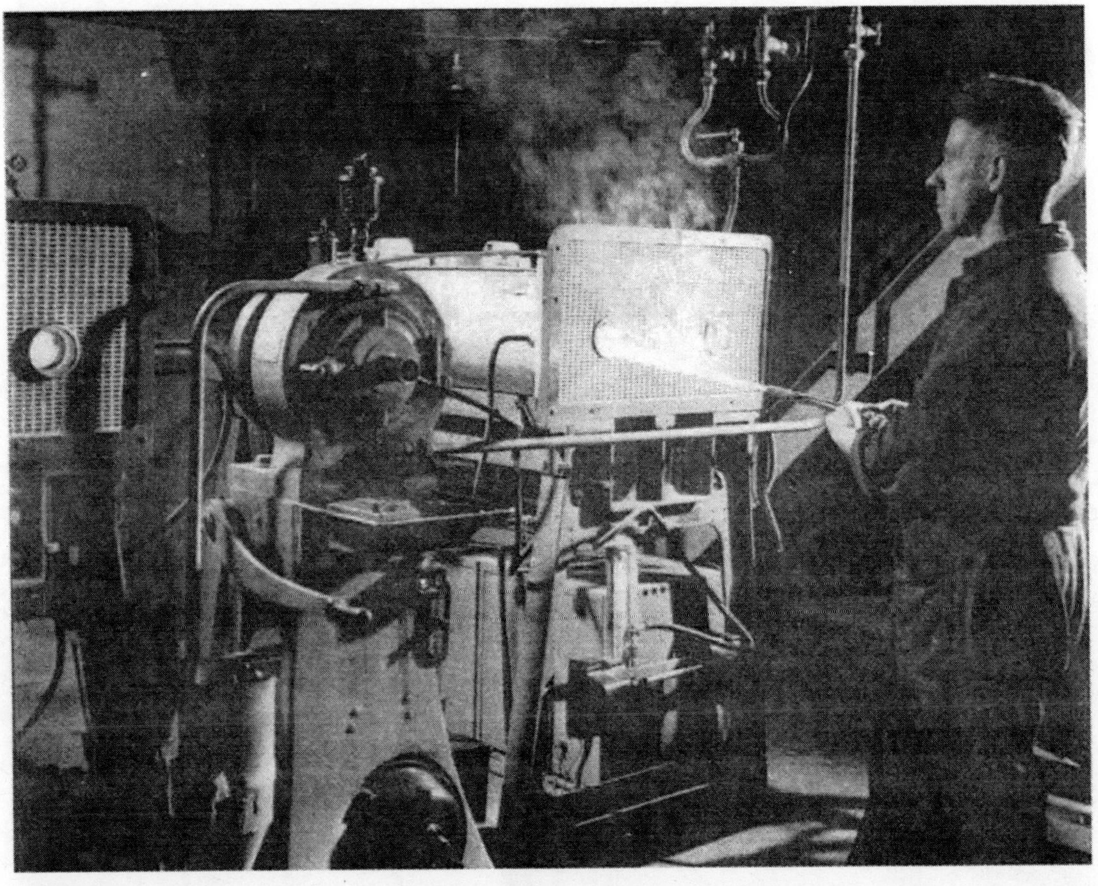

Photo 6.1. Swaging of tungsten for 'filament making'.

Ductile tungsten wire is made by machine hammering (Swaging) sintered tungsten bars to increase their density after which they are drawn to the required thickness through hard metal or diamond dies. The photograph shows a swaging machine, with the tungsten bar being introduced manually after having first been raised to a very high temperature in the adjoining furnace; after some passes through this machine, the bar is passed through an automatic Swaging machine.

(Courtesy : Philips Tech. Review, Holland).

Photo 6.2

(*a*) Structure of a wire of pure tungsten after recrystallization.

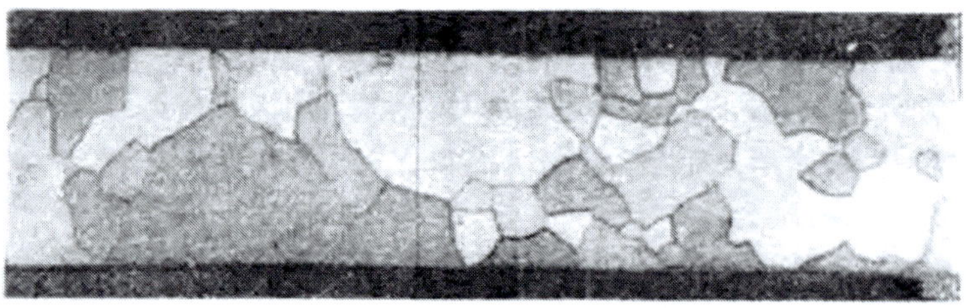

(*b*) Recrystallization texture of a tungsten wire containing impurities resulting from Battersea process.

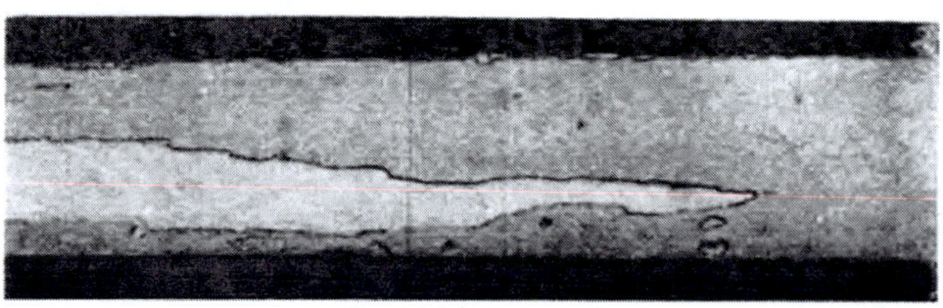

(*c*) Recrystallization texture of "Doped Tungsten" *i.e.*, tungsten containing controlled amount of additives. The elongated wedge shaped crystals give the wire considerable strength and prevent sagging when the wire is coiled.
(Courtesy : Philips Tech. Review, Holland).

Photo 6.3. A ribbon bulb shell blowing machine. The forehead 'A', blowhead conveyor 'B' and parison cooling sprays 'C'.

A fast moving ribbon of molten glass is rolled at a controlled speed between rollers where thickened areas of glass are formed and fed on to the ribbon conveyor. 72 blowing nozzles gradually indent these thickened areas on the ribbon and, together with 48 opening closing moulds, continue the blowing process until the bulbs are formed.

(Courtesy : Badalex, England).

of nickel wire in diameter from 0.40 metre to 1.00 mm are used. For vacuum lamps, oxygen free (poxidised) or OFHC (oxygen free, high conductivity) copper wire is used.

Inner Lead Shapes

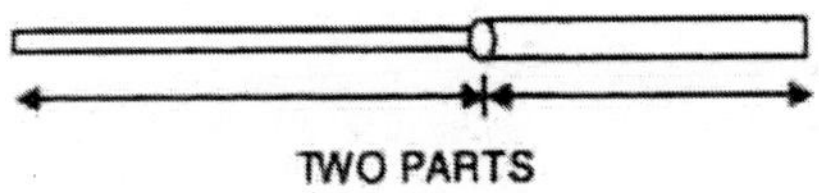

Fig. 6.4. Round shape (rd).

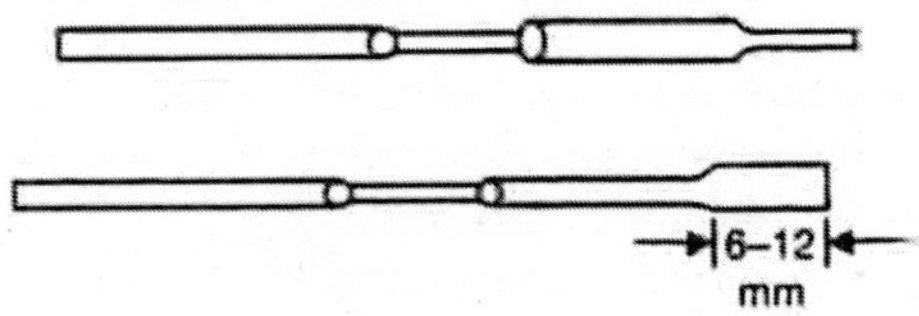

Fig. 6.5. Flat ends (Fl).

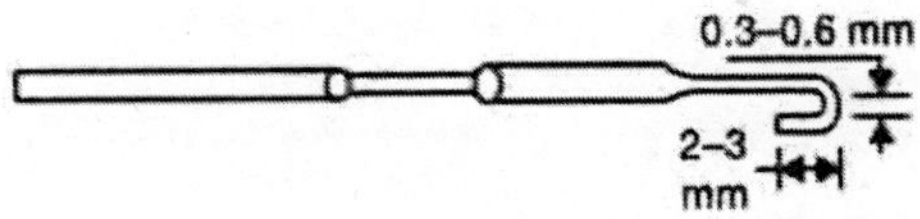

Fig. 6.6. Hook (hk).

Some Specifications of Lead-in-Wires

For Miniature Lamps — Copper clad, vacon dumet wire 0.25 mm to 0.80 mm to any desired length.

Copper clad

For Auto Lamps

Cu 0.4/40, Dumet 0.3/2.5, Ni 0.15/15 *rd*

Cu 0.5/40, Dumet 0.4/3, Ni 0.8/22 *rd*

Cu 0.5/60, Dumet 0.5/3, Ni 1.1/27.5 *rd.*

For GLS Lamps

Candle lamp—Cu 0.35/50, Dumet 0.3/4, Nipla 0.4/14–16

60 Watt—Cu 0.35/60, Dumet 0.3/4, Nipla 0.5/22–24

100 Watt—Monel 0.16/37, Dumet 0.3/15, Ni 0.6/30

1.9 mm × 23 mm glass sleeved for coiled coil lamps.

Fig. 6.7

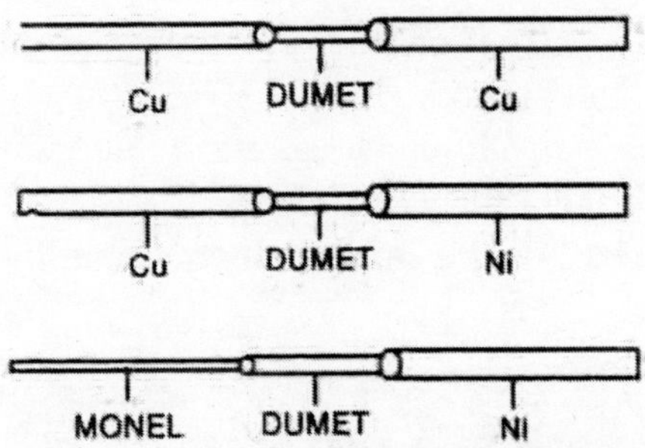

Fig. 6.8

Cu Monel Dumet Nipla

Cu 0.4/32.5, Monel 0.125/16.5, Dumet 0.35/15.5, Ni 0.6/30

Cu 0.4/36, Monel 0.16/14, Dumet 0.3/14, Nipla 0.5/26

Cu 0.4/36, Monel 0.2/14, Dumet 0.3/14, Nipla 0.6/26.

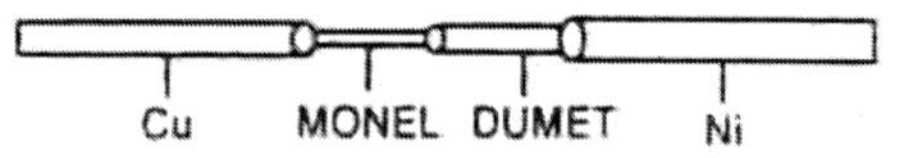

Fig. 6.9

C. GLASS IN LAMP INDUSTRY

In lamp making, by volume or weight the mostly used material is glass. In fact, the cost of production and its efficiency or productivity lies in its efficient use. In the functioning of an electric lamp the glass does not play any role but stands guard to protect all the operations. For this reason glass gets a secondary or stepmotherly treatment from workers and management alike.

Practically 40% to 50% defects in lamp making are related to glass. From handling of glass in raw material stage to the final despatch from the factory and ultimate use by the customers, it is glass alone which needs vigilant attention all the time. One should remember that glass is the only material which carries, protects and finally expresses all the technology of lamp making.

The common functions and qualities expected of the glass are :

(*a*) It serves as an isolating barrier between the inner atmosphere and the outer atmosphere of the lamp.

(*b*) The glass in clear-lamp, should transit all the light from the filament without any obstruction. Whereas in 'day light' inner coated, or etched glass tubes, the glass transmits the desired spectrum.

(*c*) It also should be able to withstand the pressure of the filling-gas, the difference of pressure of the filling-gas; the rise and fall of temperature inside the lamp.

(*d*) It should be able to withstand differences in weather conditions as cold blizzard, heavy showers or highly heated atmosphere in a casting factory shed.

(*e*) Though the glass acts as a "safe keeper" of lamp technology it never takes part in lamp operation meaning it is chemically non-reactive to any of the inner materials of the lamp.

(*f*) It should also have the quality to remain resistive to flow of electricity at higher temperature.

(*g*) It also must make and maintain for a longer period the hermetic seal at all pressure, temperature variations.

In the lamp industry, the following types of glasses are used :

Glass in Lamp Industry

Soda Lime Glass	*Lead Glass*	*Borosilicate Glass*	*Fused Silica or Quartz Glass*
GLS-Mini-Auto shell	Glass tubings and solid rods for stem	Photoflood, sealed beam, infra-red, high pressure lamp, P_2O_5 purification towers	Halogen lamps

In ancient times, the Egyptians made glass by melting natural soda with sand. Actually the soft glass is a compound named sodium calcium silicate (Na_2O CaO_6 SiO_2).

Glass is a very special supra cooled liquid of fused silicate. It is colloidal in nature and has very high viscosity. Because of this special quality it can be manipulated very easily through blowing, bending or drawing out. So for its different compositions, different temperature ranges are achieved to make different types of items. For this reason, one should know more about the 'annealing point' of the glass rather than the 'Softening Point' of the glass.

A little detail about the properties will be discussed here as the lamp maker has to take some important decisions on the types of glass to be used from studying the technical literatures of the glass makers.

Glass is a very poor conductor of heat. If a hot piece of glass is suddenly cooled it will crack and break. The reason being that the heat from the cooled zone very quickly passes to the cooling media but the heat at the opposite end still remains. This creates "stress" on the glass. Normal soda lime molecular structures cannot maintain their cohesion and the stress divides the zones apart. A piece of hot glass tubing can be prevented from undue stresses by gradual lowering down its temperature till the molecular boundaries, set themselves systematically and the undue setting up of "stresses" do not take place. This slow process of lowering down of the temperature technically known as annealing is spreaded over a wider range. Once this temperature falls below the lower limit of the range, cooling can be rapid and it will not cause any crack or breakage to the glass.

Annealing Point. Any strain caused to the glass by previous working would disappear within about 15 minutes if this temperature is achieved. The annealing point of the glass is about $10^{-13.4}$ poises.

Strain Point. This is the temperature at which the viscosity of glass is about $10^{14.6}$ poises. At this temperature any strain built up earlier in the glass will disappear within a few hours.

Transformation Point. A strainless rod is first heated at a rate of 1½°C per minute to about 600°C and then cooled at a rate of 2°C per minute and the cooling measurements are plotted on a graph. The 'transformation point' is the temperature corresponding to the intersection of two straight lines that can be drawn along the straight sections of the expansion curve on either side of the angle of the curve plotted earlier.

American Softening Point. This is the temperature when the viscosity of glass is about $10^{7.6}$ poises.

Conductivity. The sodium content in the glass makes it less and less resistive to electric current with the rise in temperature. The conductivity of glass is expressed as the temperature (*tp*) at which the resistivity of glass is $10^{6.52}$ ohms per cm.

Co-efficient of Expansion. In glass the co-efficient of linear expansion is expressed by the letter α (alpha). For measuring this property also a strainless rod is heated from 30°C per minute. The co-efficient is obtained by dividing the elongation by the length and the temperature.

Soda Lime Glass. Soda glass as commonly called is used as envelope for all types of incandescent lamps. Because of its low melting point and easy workability, this is extensively in use. The colour of the glass turns brown in the flame so it is not suitable for laboratory work.

Properties :

Specific gravity—2.55

Co-efficient of expansion—30–300°C = 102×10^{-7}

30–450°C = 106×10^{-7}

30–500°C = 112×10^{-7}

Transformation point ...490°C

Strain point ...480°C

Annealing point ...505°C

American Softening Point = 680°C

Conductivity—*tp* = 250°C

Composition. This varies with different manufacturers but a rough percentage is given here

SiO_2 — 68%

K_2O — 2%

Na_2O — 18%

CaO — 6%

Al_2O_3 — 4%

MgO — 1%

B_2O_3 — 2%

In India glass manufacturing is concentrated at Firozabad in Uttar Pradesh. There are a few manufacturers here making glass shells by mouth blowing process. Here the same soda lime silicate compositions are continuously melted in the gas fired furnaces. A blower with the aid from a helper blows the molten composition at the tip of the 'blow pipe' inside a die. The strained shell is then placed inside the Annealing Furnace for relieving it of all the strains and stresses.

The large manufacturers have their own arrangements of both mouth blown and automatic machines for manufacturing lamp shells.

In automatic high speed shell blowing machine, the glass flows at a pre-determined rate from the continuously fired furnace between sets of water cooled rollers. (*See Photo 6.3*). These rollers form very thin 'ribbons and rounds' at distance of about 100 mm from one another. The ribbon is then picked up on to the next transfer chain and the round portion falls over a hole in the die. The die here is blown, continuously rotated and also sprayed with cool water which finally gives a lusterous polished outer surface of the shell. The ribbon or the holding tag is broken away after removing the die and the completed shell is then put on to the conveyor proceeding to the annealing furnace.

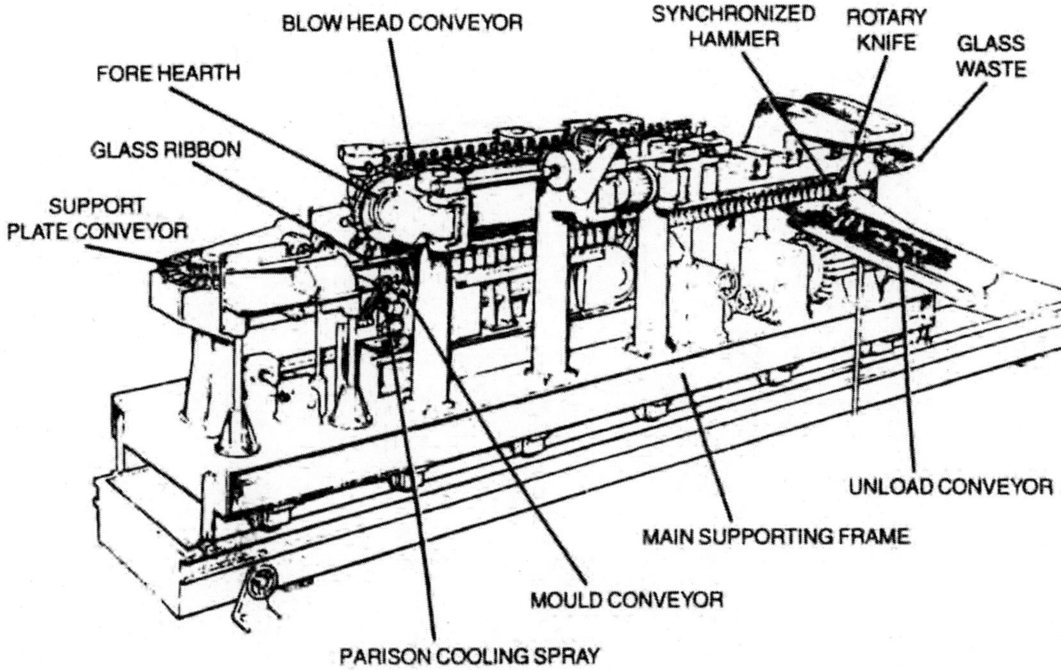

Fig. 6.10. Ribbon bulb blowing machine.

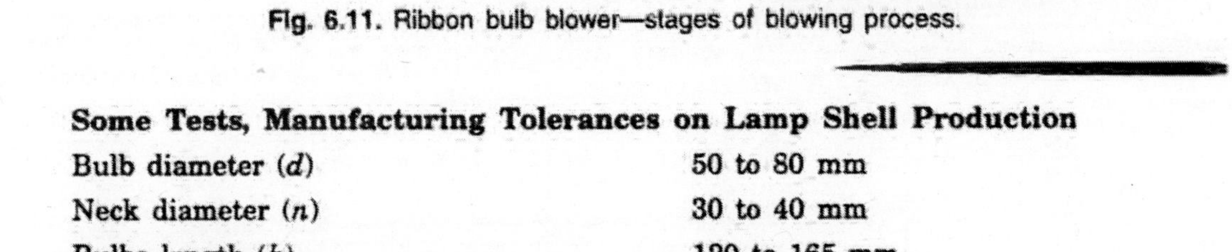

Fig. 6.11. Ribbon bulb blower—stages of blowing process.

Some Tests, Manufacturing Tolerances on Lamp Shell Production

Bulb diameter (*d*)	50 to 80 mm
Neck diameter (*n*)	30 to 40 mm
Bulbs length (*h*)	120 to 165 mm
Average wall thickness in sealing zone (*W*)	0.55 mm
Minimum curvature of radius	10 m/m
Length (*h*)	250 m/m
Tolerance upto 250 m/m	2.5 mm.

The shells are visually inspected of critical defects, major defects and minor defects.

Some Tests Manufacturing Tolerances on Lamp Shell Production

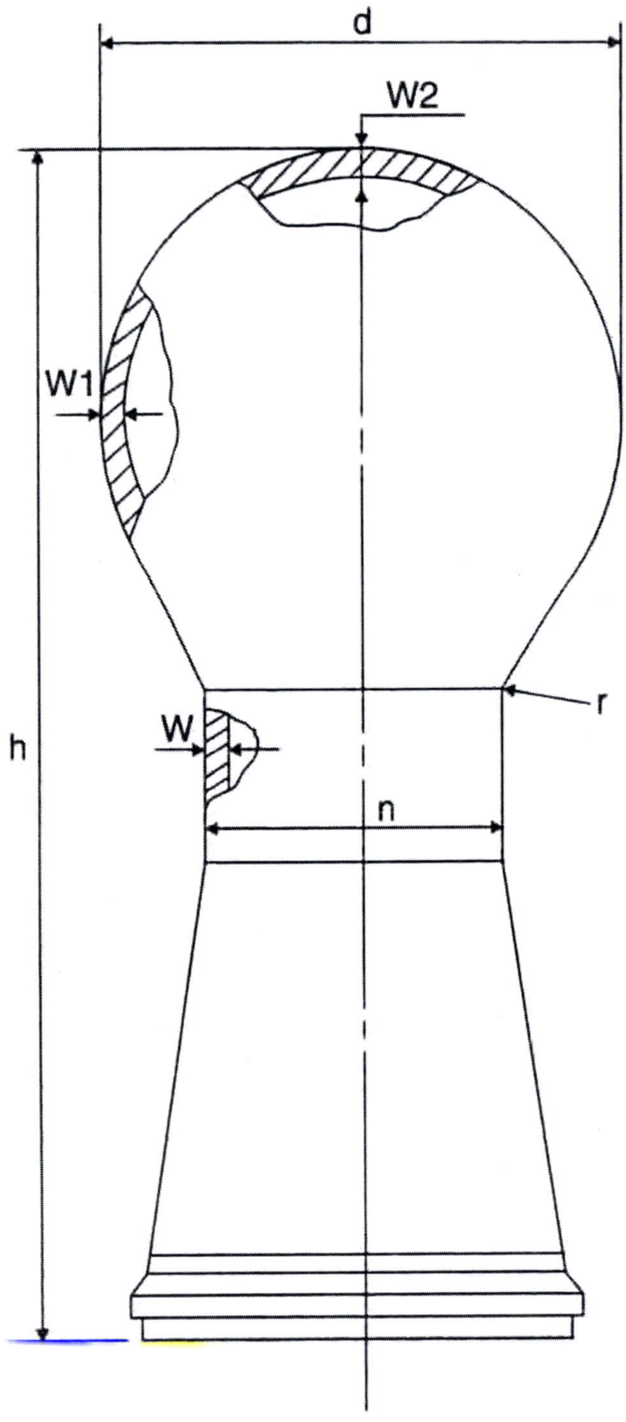

Fig. 6.12

Some Critical Defects

(*a*) Extra sticking glass piece

(*b*) Crack on sealing area

Major Defects

Crack under sealing area, twisted neck, open bubbles.

Minor Defects

Rough bowl, bubble, dents, stones and knots scratches etc.

The shell samples are tested on compression strength testing apparatus and the shell should withstand a pressure of 8 kg.

Lead Glass. Lead alkali silicate glass is specially used in lamp industry for manufacturing lead glass tubes and solid rods. At higher temperatures, the resistivity of Soda lime glass recedes to a large extent and electrolysis action takes place between the lead-in-wires inside the soda lime stem pinch. The lead alkali silicate has higher resistivity at higher operating temperature. Moreover lead glass easily combines with soda lime glass shell, it has a lower working temperature and longer working temperature range.

A typical composition of Lead glass tubing is as following :

		Properties
PbO	30%	Specific Gravity = 3.05
Al_2O_3	1%	**Co-efficient of expansion**
SiO_2	56%	(alpha) α 30–406°C = 95 × 10^{-7}
Na_2O	5%	(alpha) α 30–450°C = 106 × 10^{-7}
K_2O	80%	Transformation point .... 425°C
		Strain point .... 400°C
		Annealing point .... 425°C
		American softening point .... 625°C
		Electrical conductivity......................... 365°C

Production. At Firozabad, lead glass tubing is mouth blown and hand drawn. By this manual process it is never possible to get uniformity either in outer diameter or wall thickness.

In the automatic machines, it is made by Daner process. The molten glass from the furnace is sent on to the top of a hollow refractory mandrel through which compressed air is blown. The end of the molten glass which creeps down the mandrel is pulled to a 40–50 metres distance by a drawing machine at a controlled uniform speed. Further the glass moves away from the mandrel it gets cooler and cooler solidifying as a smooth uniform dia lead glass tubing. The general defects in glass tubes and rods are cracks, chipped surface, long air lines, excessive hardness, visible impurities in composition etc.

In the market lead glass materials are available under the following specifications :

	Outer Diameter	*Wall Thickness*	*Overall Length*
1. Flange tubes	6.0–6.5 m/m to 17.5 to 18 mm	0.9 to 1.7 mm	1100 mm
2. Exhaust tubes	1 mm to 6 mm	0.6 to 0.9 m/m	1140 m/m
3. Solid rods	3.25 to 3.75 m/m	—	1140 m/m

The colour of glass tubing is generally transparent and colourless but may take a light violet tinge which is acceptable. Lead glass should be always worked with the aid of oxygen and compressed air. Any work carried with lead glass is to be followed with gradual reduction of temperature in an Annealer.

BOROSILICATE GLASS

Borosilicate glass is commonly known by their trade name as corning 7740, phoenix, pyrex or GTC–B 37. All the above types are "hard glass" which can withstand thermal shocks without fracture and high temperature work without deforming. These have optimum resistance to mechanical shocks and resist chemical attacks.

In addition to its abilities to withstand higher operating temperature, the borosilicate glass has a very low co-efficient of expansion which makes it possible to withstand greater changes in temperature.

This glass is mainly used in the lamp industry for projector lamp shells, high wattage stems, sealed beams, high pressure discharge lamps and some very small mini lamps where accuracy of shape is of utmost importance. To withstand both high pressure and vacuum and also remain non reactive to P_2O_5 derivatives, this borosilicate glass is used in making purifying towers for argon-nitrogen gas.

It is a low alkali borosilicate glass and is virtually free of magnesia lime-zinc group and contain only traces of heavy metals. It's typical chemical composition and properties are as following :

Composition %	*Pyrex*	*Corning 7740*	*GTC B 37*
SiO_2	80.8	81	76
B_2O_3	12.0	13	16
Al_2O_3	2.2		
Fe_2O_3	0.05		
Na_2O	4.4	4	4
AS_2O_3	0.05	2	2
K_2O	—	—	2

Properties	*Pyrex*	*Corning 7740*	*GTC B 37*
Co-efficient of linear expansion	32×10^{-6}	32×10^{-7}	32.5×10^{-7}/°C
Straining point	—	—	515°C
Annealing point	—	525°C	565°C
Softening point	—	775°C	820°C

Borosilicate glass always should be worked with oxygen enriched flame. If the glass is heated above 500°C, the glass may acquire permanent stress on cooling.

Fuses Silica or Quartz Glass

This is explained in Chapter 8 under glasses for halogen lamps.

PART II

PRACTICE

If one knows—
'Why' an operation is to be made ;
the how-where-when then
depends on his choice.

7

MANUFACTURING OF INCANDESCENT LAMPS

A. TRADE CATEGORISATION

Depending on their applications, there are more than 500 types of electric lamps available in India. Taking the diameter of the shell as the basis for differentiation, the Indian Standards Institute has devided the lamps in the following three categories.

Category 1. Miniature lamps or mini lamps in short. The small lamps having shell diameter from 6 mm to 18 mm fall under this category. Normally the 'festoon lamps' or the decoration lamps are known as mini lamps but these are not yet brought under the perview of the Indian Standards. For automobiles, the side lamps, dash-board lamps, tail lamps come under this category. All torch lamps, indicator lamps, panel lamps also fall into this group.

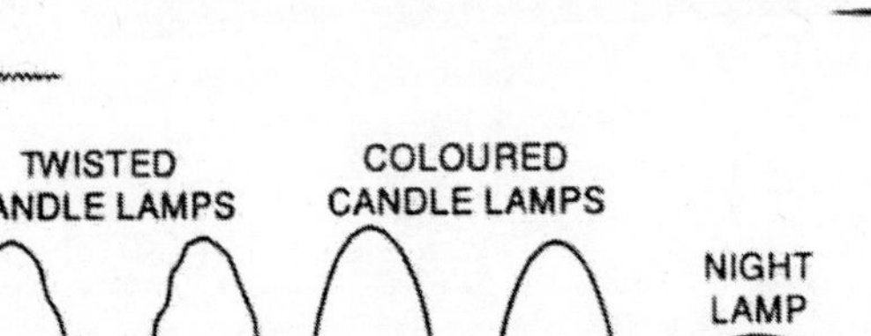

Fig. 7.1. Figures and shapes of some miniature lamps.

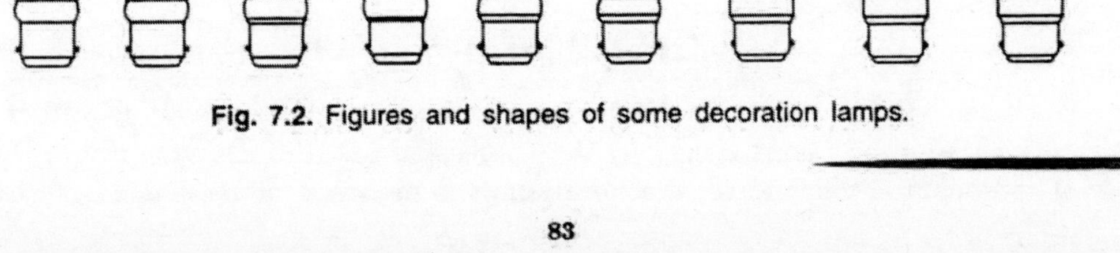

Fig. 7.2. Figures and shapes of some decoration lamps.

Category 2. All the lamps between 18 mm and 35 mm diameter fall under this category. As almost all the lamps made in this range cater to the automobile lighting, this group is generally termed as 'Auto Lamps'.

For catering to the needs of the auto industry, the auto lamp makers also manufacture lamps from mini category, which are also termed as auto lamps by the trades people—the difference being that the mini auto lamps also are standardised by the I.S.I.

Category 3. All lamps having shell diameter from 45 mm and above fall under this category. The lamps used for providing light in general are known as General Lighting Service Lamps (GLS in short). The coloured, frosted, or clear lamps of smaller wattage, the candles, pear shaped decorative lamps and all the lamps which do not fall into category 1 and 2, are taken to be GLS lamps which need be clearly defined as yet.

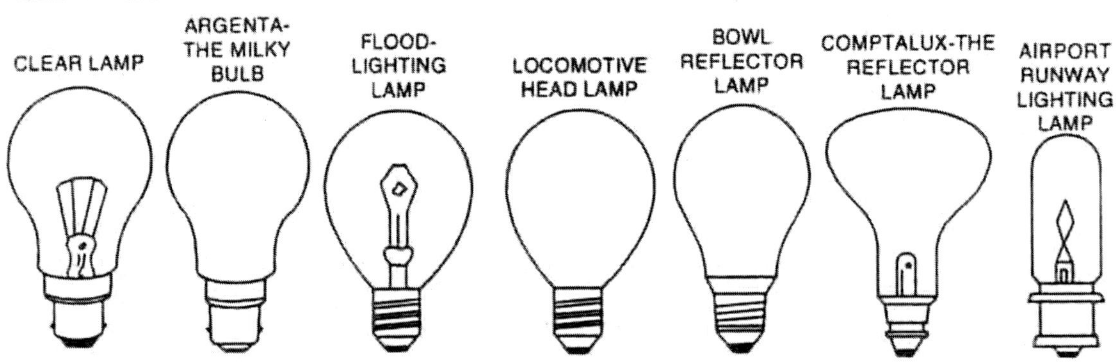

Fig. 7.3. Figures and shapes of some GLS and special lamps.

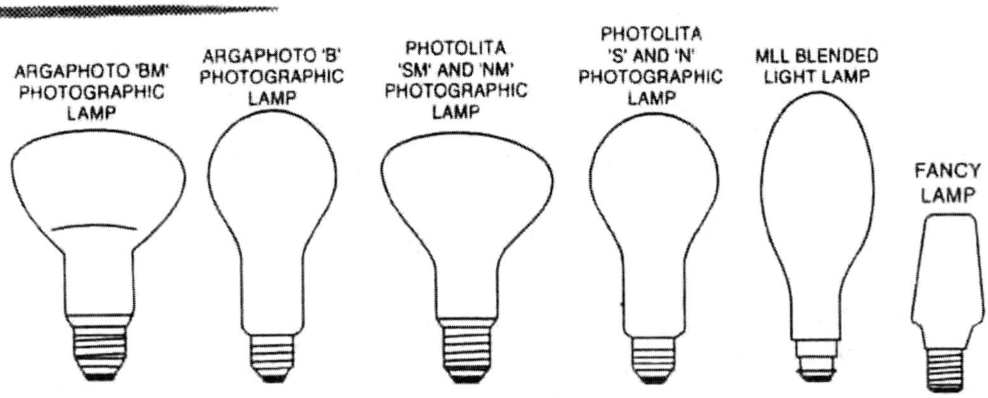

Fig. 7.4. Figures and shapes of some special lamps.

1. THE MINIATURE LAMPS

The term miniature does not have any technical significance. It is rather a name indicating small sized lamps coined by the tradesmen and the manufacturers. During the festival seasons, the demand for the mini lamps is high and many small factories engage themselves in manufacturing 'festoon lamps'. These small festoon lamps are made in millions in the small scale or tiny scale factories and much attention has not yet been directed towards its technical qualities as their services are necessary only for the short period of the festival. It is hoped that in due course these lamps also will be standardised and the users will have the prior knowledge about the characteristics of these lamps. So, while speaking about mini lamps, these should not be misunderstood as 'festoon lamps'.

In most of the cases, the mini lamps are used as car signal lamps, panel lamps, in light oriented appliances as flash light, instrument panel lights, cabin lights, surgical instrument lamps etc.

Mini lamps are made for particular purpose and their use with particular type of power source *i.e.,* dry cell, lead acid accumulators, or transformers. The mini lamp should be designed to withstand the mean effective voltage of the power source which is always greater than the available average voltage.

Being so small in size, the light and life of the mini lamps are so related that to get a better efficiency of one, the others quality is to be sacrificed. With the internal space available inside the mini lamp, it has been found that the range between 6 volt to 24 volt is the most suitable and efficient power supply system. At voltage below 6 volt, the lead-in-wires carry outside the heat of the filament to the metallic cap. When a 5 volt flash light lamp has an efficiency of 1 candle power per watt, a 2.5 volt flash light has 0.87 candle per watt and a 1.25 watt flash lamp has only 0.25 candle power per watt. It shows clearly how the heat has reduced ultimately the effective lighting power of the lamp.

Keeping the size constant, the life and light factor of the lamps is dependent on the applied voltage. The voltage being the electrical pressure, may be termed as the main contributor to the life of the mini lamp. The light out put of the lamp varies as to the 3.6th power of the voltage while the life varies inversely to the twelfth power. If a 6 volt lamp is burnt at 5 volt, the light efficiency will drop to it's half whereas the life of the lamp will become ten times more. This is one of the reason to select 48 number to make a chain of festoon lamps. These are all 6 volt 'festoon lamps' (generally 6/2 type Group–I) and are connected in series with domestic power supply of 220–240 volts. Even at peak voltage each of these lamps shall never receive more than 5 volts. The lamps are connected nearer to one another with coloured covers so that their luminance is never in question and at the same time these get a longer span of life too.

Utmost care must be taken to manufacture 6 volt radio panel lamps. The radio chassis is always under constant shock and vibrations. There is a general complaint that these lamps fail prematuredly. The reason being that the filament coil of these lamps do not resonate at the same frequency of the chassis. The manufacturers of the radios and the panel lamps must work in unision and the former should inform of the resonance frequency of the chassis to the latter. The lamp maker has to then select the suitable filament coil which would resonate at the same frequency of the chassis. Practically a lamp made in such collaboration lasts for many years.

For increasing the life of a 6 volt lamp, inert gas also is introduced which retards tungsten carriage to the shell wall. It is not necessary to fill it with expensive argon-nitrogen mixture. The voltage being so small, there is little chance of internal sparking in these lamps and pure dry nitrogen also serves the purpose.

The cement and the adhesive glue used for capping retain their properties upto 175°C. If a lamp is used at a place where the working temperature is higher, the butt sealed bare type lamps are recommended to be used.

The mini lamps must not be used at places where the ambient temperature is 300°C or above as the shell will collapse at this temperature.

(*See Photo 7.1*)

In Indian Standards, the small lamps with a maximum shell diameter of 6.8 mm to 18 mm and maximum length of 17 mm to 35 mm are known as miniature lamps (with prefix reference no. 'M'). The mini lamps are put into five groups as per their size and shape of their shells, for easier identification as mentioned in the following page.

Type I.S. Ref. No.	*Volt*	*Watt*	*Cap*	*Lumen / Watt*	*Life in Hours*
MINI SIZE—(Shell dia 6.8 mm and length 17 mm)					
*M*6/1	6	0.6	*BA* 7s	5.0	300
*M*6/2	6	1.5	*BA* 7s	5.6	300
*M*12/1	12	1.2	*BA* 7s	5.0	300
*M*24/1	24	1.5	*BA* 7s	5.3	400
MINI SIZE—(Shell dia 8.5 mm and length 23 mm)					
*M*12/2	12	2.2	*BA* 9s	5.6	400
*M*24/2	24	1.5	*BA* 9s	5.3	400
MINI SIZE—(Shell dia 11 mm and length 24 mm)					
*M*6/3	6	1.5	*BA* 9s	5.6	300
*M*6/4	6	3	*BA* 9s	6.4	400
*M*12/3	12	2.2	*E* 10	5.6	400
MINI SIZE—(Shell dia 15 mm and length 29 mm)					
*M*12/5	12	6	*BA* 9s	7.2	100
*M*12/4	12	6	*E* 10	7.2	100
MINI SIZE—(Shell dia 18 mm and length 35 mm)					
*M*12/6	12	6	*BA* 15s	7.2	100
*M*24/3	24	6	*BA* 15s	6.6	200
*M*6/5	6	3	*BA* 15s	7.5	200

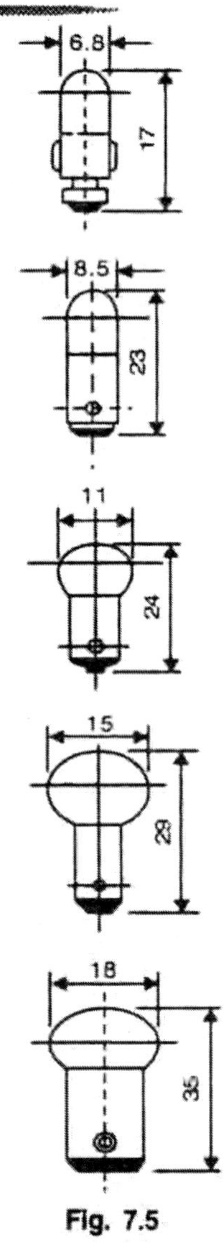

Fig. 7.5

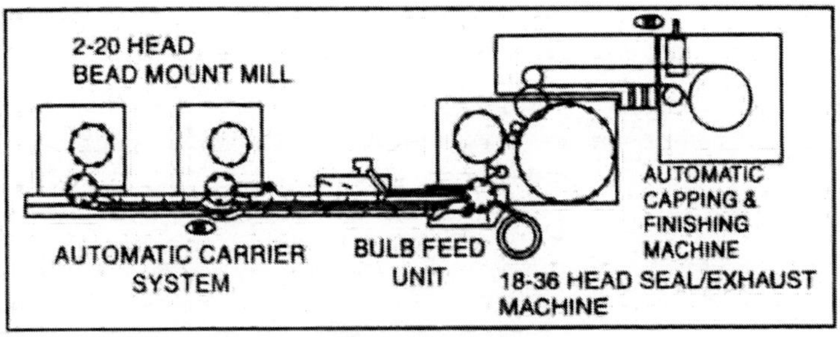

Fig. 7.6. Layout of automatic transfer chain for butt-seal miniature lamp.
(Courtesy : Badalex—A division of Sale Tinley Technology, England)

2. THE AUTOMOBILE LAMPS

The electrical system in an automobile is fed by a lead acid battery either of 6 volt, 12 volt or 24 volt capacity. This battery is constantly charged by a generator run by the engine. The charging takes place through a cutout system which separates the generator from the charging circuit at low revolutions of the engine and at high speed the generator directly feeds the electrical system. The capacity of the battery is limited but it has got to provide the initial starting heavy current for the starter, power to the running engine and at the same time provide power to all the head lights, tail lights, side lights, cab lights, signal lights and panel lights. It is also imperative that at all road conditions *i.e.*, broken roads, cross country, extreme weather conditions, rainy seasons etc., all the lights should continue to be operative. In the city areas, the battery and the generator supplying the current should be in perfect condition to get proper light. Above all the jerks on the starting, stopping and acceleration call for perfect and high skilled worksmanship in manufacturing an auto lamp. The constant changes in the voltage supply with chattering sparks from the cutout system, the sparks from the generator brushes are all constant faulty companions of the automobile system. The makers of the auto lamps are to take note of these inherent problems and at the same time the users of the automobiles have to keep their electrical system in good condition to get the best service from their lamps. An auto lamp is mostly a mini lamp having all the characteristics of a GLS lamp and for certain manufacturing operations, it is even more difficult and sophisticated than producing a GLS lamp.

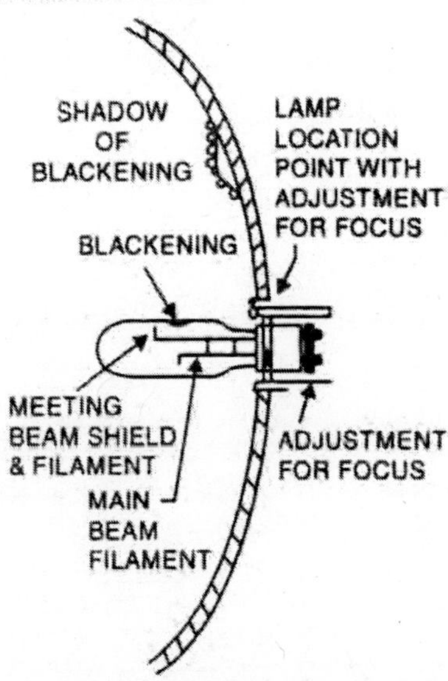

Fig. 7.7. Prefocus auto lamp with reflector installation.

Broadly the auto lamps can be divided into the following three groups.

Group I. The small lamps having shell diameter upto 18 mm as described earlier under Miniature Lamps, fall in this group.

Group II. (Shell diameter 25 mm and lamp length 79 mm).

With limited source of power supply, limited space for installation, these lamps are to provide enough light, compactness and rigidity. These auto lamp do not require a pre-focussed centre. The caps have single point (marked *s*) or double points (marked *d*) as per their place of use. Proper light centres are to be maintained in these lamps.

Type I.S. Ref.	*Volt*	*Watt*	*Cap*	*Lumen per watt*	*Light centre ± 1.5 m/m*	*Life in hours*
Interior light etc.						
D 12/2	12	21	*BA* 15 *s*	15.7	31.8	40
D 24/2	24	24	*BA* 15 *s*	15.7	31.8	30
Stop light etc.						
D 12/1	12	12	*BA* 15 *d*	13.2	31.8	1000
D 24/1	24	20	*BA* 15 *d*	8.2	31.8	1000
Stop and tail lights with two mountings						
D 6	6	6/18	*BAY* 15 *d*	4.5	31.8	500
				14.1		30
D 12/3	12	6/21	*BAY* 15 *d*	4.00	31.8	500
						40

Figure

Fig. 7.8

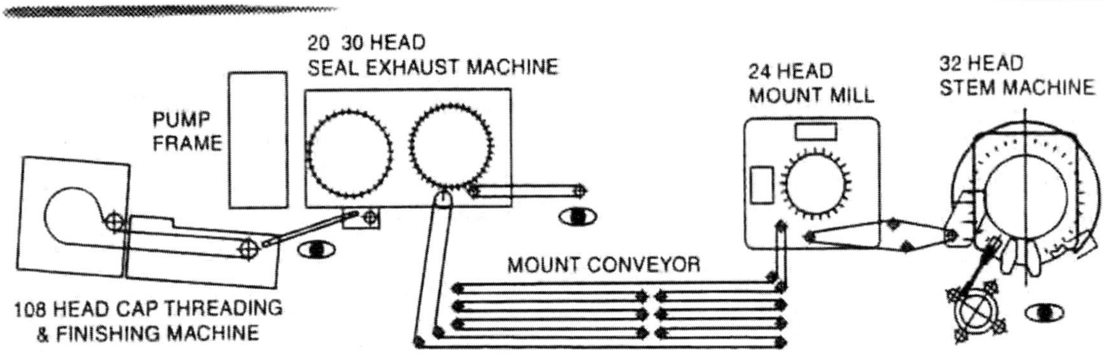

Fig. 7.9. Layout for automatic automobile tail/stop lamp manufacturing transfer chain.

Production rate. Fully automatic—2,2000 lamps/hr.
Semi automatic—1500 lamps/hr.
(Courtesy. Badalex—A div. of Sale Tinley Technology—England).

Group III. (Shell dia 28 mm to 41 mm and maximum permissible lamp length is 70 mm).

All the Head Lights come under this group.

In head lamps the amount of light aimed at a particular angle and spot is of greatest importance. Normally these have two filaments, one for the main beam giving maximum light for better visibility of the road and the road sides and the other filament is used with less wattage for meeting beam known as 'dipper light'. The beam of the head lamp is concentrated and then focussed by means of reflectors. The automobile manufacturers use different types of head lamps with various types of reflectors to suit their purpose. In some rough running and heavy vehicles, 'Sealed Beam' head lamps are used (See. Fig. 7.7). In most of the automobiles, replaceable lamp type head light units are preferred. The prefocussed lamps use a focussing collar which is easily distinguishable from the general type of head lamps.

Asymmetric E Type Head Lamps. As the pre-focus lamp this also has two separate

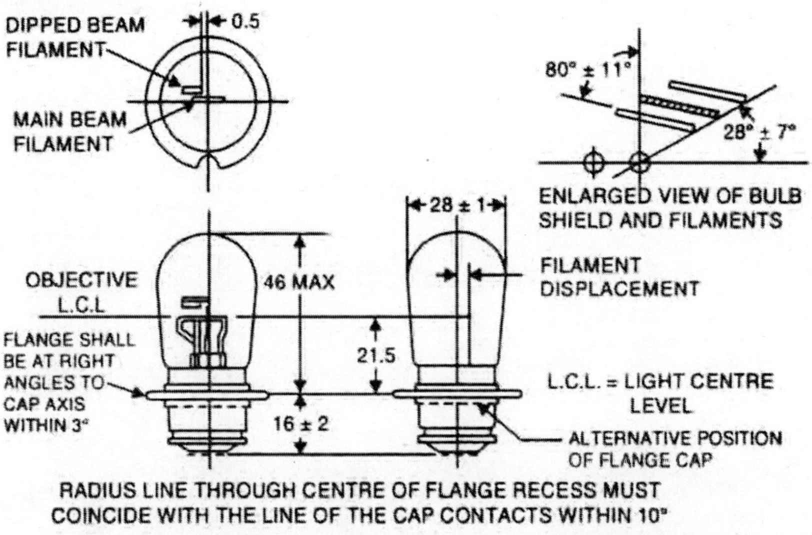

Fig. 7.10. Head light lamps (Anglo American Beam Pattern).
(Courtesy : Indian Standard Institute, India)

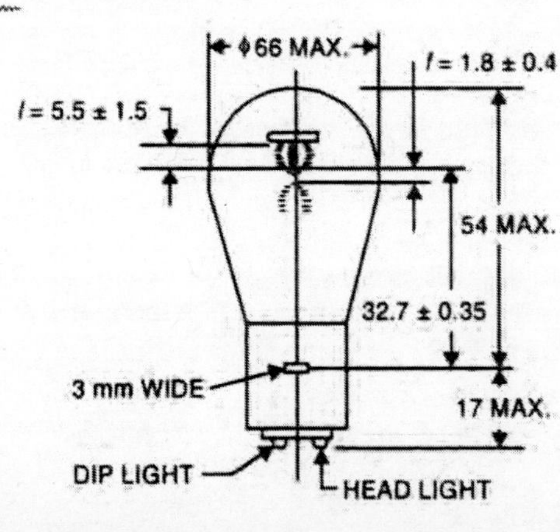

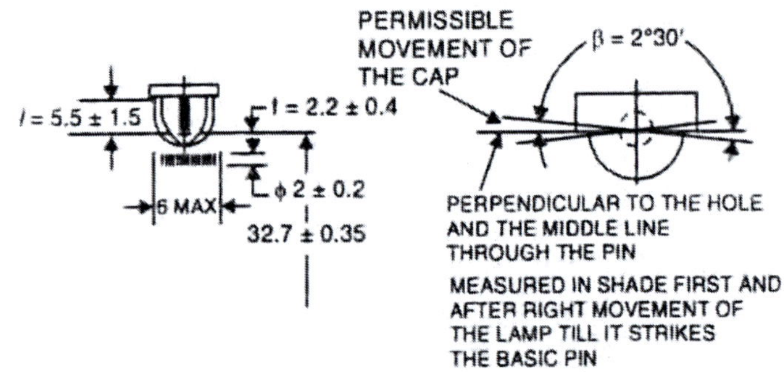

Fig. 7.11. Head light lamps (European Symmetric Beam Pattern).
(Courtesy : Indian Standard Institute, India)

filaments for main beam and meeting beam. The meeting beam is mounted in a cavity reflector which cuts off light above horizon on the driver's side by 15 degrees for right hand drive vehicles.

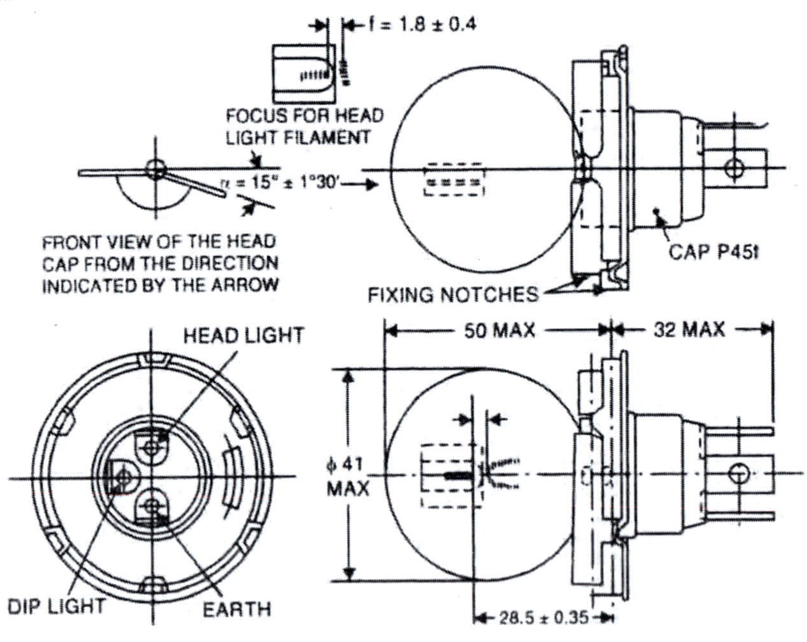

Fig. 7.12. Head light lamps (European Asymmetric Beam Pattern).
(Courtesy : Indian Standard Institute, India)

The main filament is bow shaped and is mounted at the centre of the shell. The meeting beam is mounted above it. This lamp has a straight sided cap carrying three lugs through insulator. When fitted with prefocus flange this is commonly called as 'cap on cap'. Generally 0.8 × 32 mm electrodes are used. The international rating is 12 volt 40/50 watt but in Indian Lamp Industry these are made as 12 volt 50/60 watt. As head light lamps,

Photo 7.1. Machine layout for automatic flash light lamp production.

The five fully-interlocked machines in the complete group shown in the photograph on the right are designed to index at speeds upto 2,250 an hour. They include, left to right :

(1) The Bead Mount Mill where beads and electrodes are assembled and filaments of 1.5 to 2.5 mm (coil length) automatically wound : (2) the automatic conveyorisation system : (3) the miniature seal/exhaust machine : (4) the cap filling machine, rear right, and (5) the cap threading and finishing machine, foreground right.

Photo 7.2. Automatic bulb unscrambler chain.
(Courtesy : Badalex–A Div. of Sale Tinley Technology, England.)

these Asymmetric E-types are becoming very popular and are known as new model lamps. (Figs. 7.10, 7.11 and 7.12)

Leaving aside the above mentioned asymmetric head lamps, the commonly used head lamps may be placed under the following three groups as per their sizes :

Auto Head Size–1

(Single filament, pear shape shell dia 35 mm and length 70 mm, light centre length 28.5 ± 1.25 m/hr.)

Type	*Volt*	*Watt*	*Cap*	*Lumen per watt*	*Life in hours*
H-6/1	6	24	*BA 15s*	13.2	100
H-12/1	12	12	*BA 15s*	17.4	100

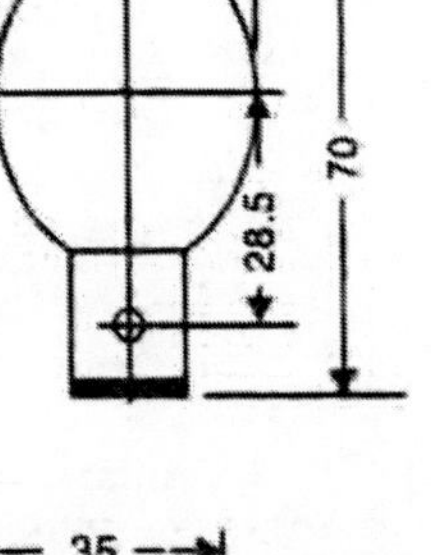

Auto Head Size–2

(Double filament, arrow head, shell dia 35 mm and length 70 mm, light centre length 31.5 + 0.35.)

Type	*Volt*	*Watt*	*Cap*	*Lumen per watt*	*Life in hours*
H-6/2	6	24/24	*BA 20d*	13.5	50
H-12/2	12	42/36	*BA 20d*	15.0	75
				14.5	100

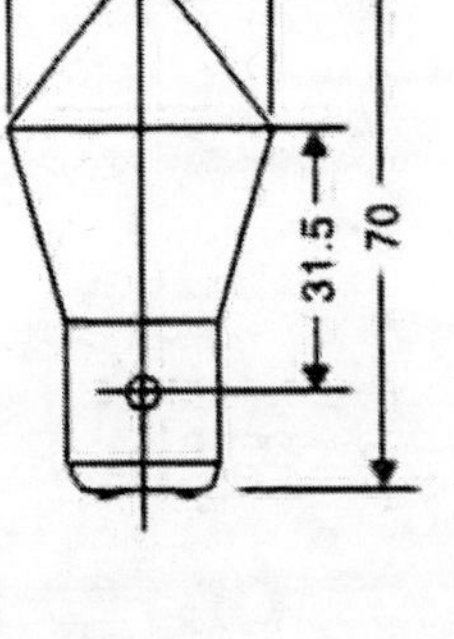

Auto Head Size–3

(Prefocus pigmy shaped, shell dia 35 mm and overall length 62 mm, light centre length 21.5)

Type	*Volt*	*Watt*	*Cap*	*Lumen per watt*	*Life in hours*
H-6/3	6	24/24	*P 2d*	13.5	50
H-12/3	12	42/36	*P 22d*	15.0	75
				14.5	100
H-24	24	42/36	*P 22d*	13.2	350
				12.4	475

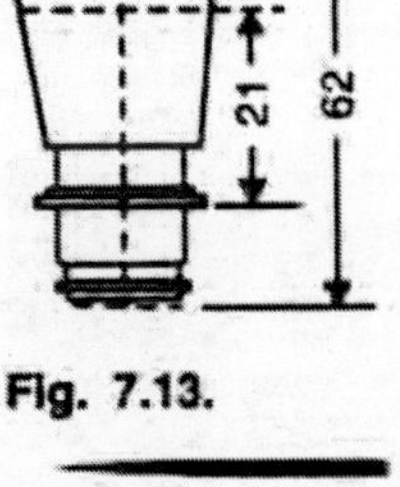

Fig. 7.13.

Some Lamps and Cap Designs

(*i*) Single and Double Filament—Head Light

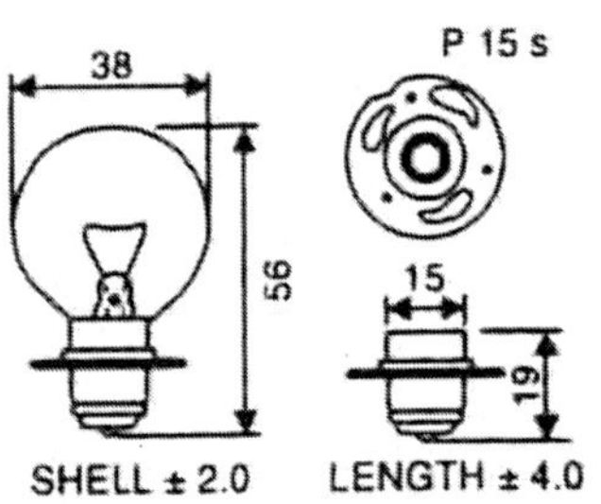

(*ii*) American Prefocus—Head Lamp (Single Filament)

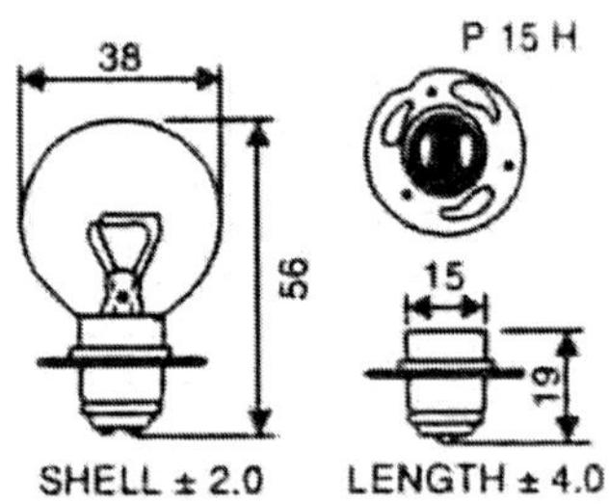

(*iii*) American Prefocus—Double Filament Head Lamp

Fig. 7.14. (*a*) Some lamp and cap designs.

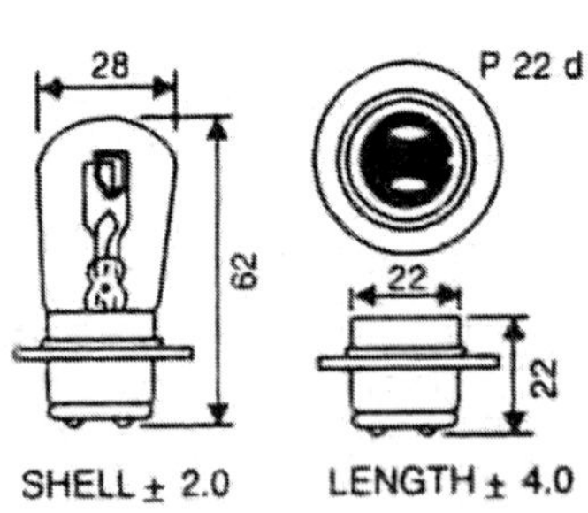

(*i*) British Prefocus—Double Filament Head Lamp

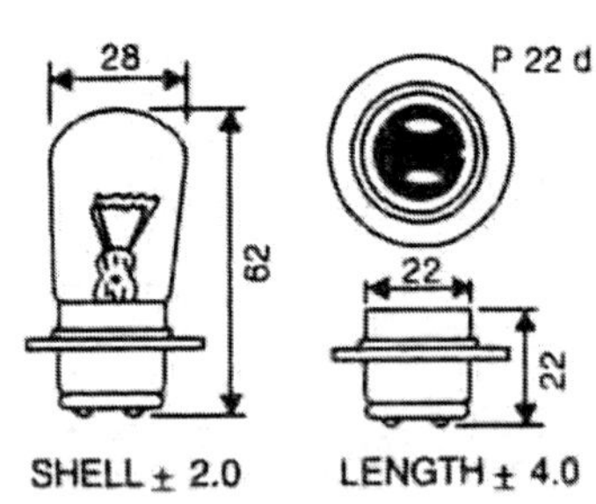

(*ii*) British Prefocus—Double Filament (with Reflector) Head Lamp.

(*iii*) Bosch Double Filament (with Reflector) Head Lamp.

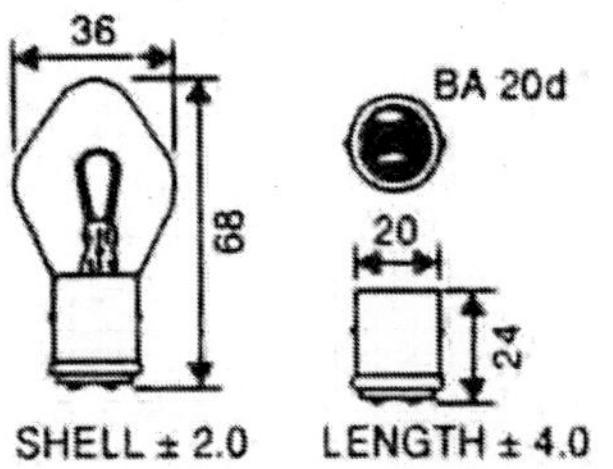

(*iv*) Bosch Double Filament (without Reflector) Head Lamp

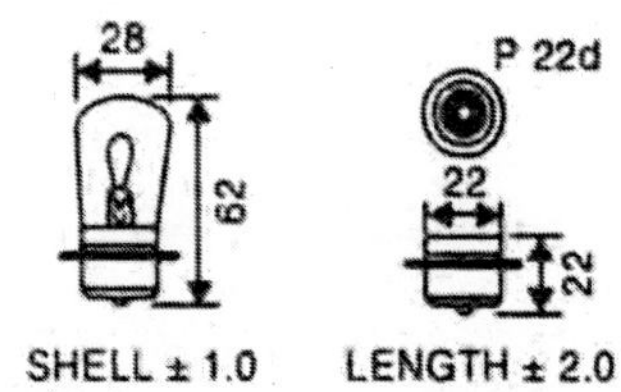

(*v*) British Prefocus Single Filament Head Lamp

(*vi*) Single and Double Filament—Stop Light.

Fig. 7.14. (*b*) Some lamps and cap designs.

3. THE GENERAL LIGHTING SERVICE (GLS) LAMPS

The general lighting service lamps are mainly used for lighting homes, offices, shops, streets etc., with the commonly accessible power supply as 110 volts (trains) 230 volts to 250 volts (domestic supply). Though more and more discharge lamps as mercury vapour lamps, sodium vapour lamps and fluorescent tubes are being brought into use, the incandescent tungsten filament lamp is still the universal light source.

The GLS lamps with a normal life of 1000 hours consist of coloured (night lamps), clear or frosted glass shell lamps and are connected to the power supply through Bayonet caps or Edison Screw Caps.

For economy of transmission the rated voltage of domestic power supply (100 volts to 250 volts) has to be kept high but its contact is fatal to human beings. So, it would be dangerous if any leakage of current would occur through any part of the GLS lamp. The first and foremost thought for the manufacturers to avoid such occurrence would be to use proper type of lamp cap with perfect insulation as mentioned in the Indian Standard (IS). Most often this item is overlooked and many lamps are found in the market with cracked, broken or substandard caps which endanger the lives of the users. For protection of the wiring from electrical fire, the lamp cap temperatures also should not exceed the prescribed range as detailed in the I.S. In the higher wattage lamps, fuse incorporated lead-in-wire are to be used. Though lighting is the most desired function of the electric lamp—the safety factor should receive the foremost attention. Under any vagaries of weather *i.e.*, hot, cold or rain, there should not be any explosion of the glass shell. The explosion of the glass shell takes place at the minute crack formations caused due to non-releiving of stress and strains of manufacturing processes. These are sometimes seen on to the neck of the lamps just at the bottom of the cap seat where stress in the form of white lines as nettings are formed due to unregulated cap fixing temperature.

For tradesmen it is adviseable not to stock all the vacuum lamps in one place. These lamps burst out violently and may cause damage to other lamps in the same stock. While testing a lamp before handing out to a customer, if the lamp gets filled with white coloured gas or shows multi-coloured markings, the testing should be stopped immediately to avoid chances of explosion. If there be metallic sound, inside a lamp, it would mean presence of foreign materials around the tail ends of the bare electrodes inside the lamp. These lamps at times blow the main fuse due to internal short circuit. The GLS lamps as manufactured in India may be categorised in the groups as mentioned in the following page.

[**For all GLS lamps.** Power supply 110 volts to 250 volts minimum life—1000 hours.]

Group I. (Low Wattage Vacuum Lamps (5 Watt to 25 Watt)

Type	*Rated wattage*	*Shell dia mm.*	*light centre*	*Filament type*	*Cap size*	*Lamp finish*
GI/A	5	45	No ESS	Single coil	B 22/25 × 26	Coloured and Frosted
GI/B	10	—do—	—do—	—do—	—do—	—do—
GI/C	15	—do—	—do—	—do—	B 22/25 × 26 E 27/27 E 27/30	—do—
GI/D	25	60	50	—do—	—do—	Clear and Frosted

Group II. (Medium Wattage—Gas Filled—25 Watt to 100 Watt)

Type	*Rated wattage*	*Shell dia mm.*	*Light centre*	*Filament type*	*Cap size*	*Lamp finish*	
GII/A	25	60	50 ± 3 mm.	coiled coil	B 22/25 × 26 E 27 × 27 E 27 × 30	Clear and frosted	Recently introduced under gas filled category
GII/B	40	—do—	—do—	—do—	—do—	—do—	
GII/C	60	—do—	—do—	—do—	—do—	—do—	
GII/D	100	—do—	—do—	—do—	—do—	—do—	

Group III. (High Wattage, Gas Filled—150 Watt to 300 Watt)

GIII/A	150	80	84 ± 4	single coil	B 22/25 × 26 E 27 × 27 × 30 E 27 × 30	Clear and frosted
GIII/B	200	—do—	—do—	—do—	—do—	—do—
GIII/C	300	90	93 ± 5	—do—	E 27 × 30	—do—
		110	128.5 ± 5	—do—	E 40 × 45	—do—

Group IV. (Extra High Wattage—Gas Filled—500 Watt to 1500 Watt)

GIV/A	500	130	128.5 ± 6	single coil	E 40 × 45	Clear
GIV/B	1000	150	156.5 ± 8	—do—	—do—	—do—
GIV/C	1500	170	173.5 ± 8	single coil	E 40 × 45	Clear

B. MANUFACTURING CATEGORISATION

The manufacturing methods take into consideration the internal construction of the lamp for which maximum efficiency in production could be obtained with easier and lesser

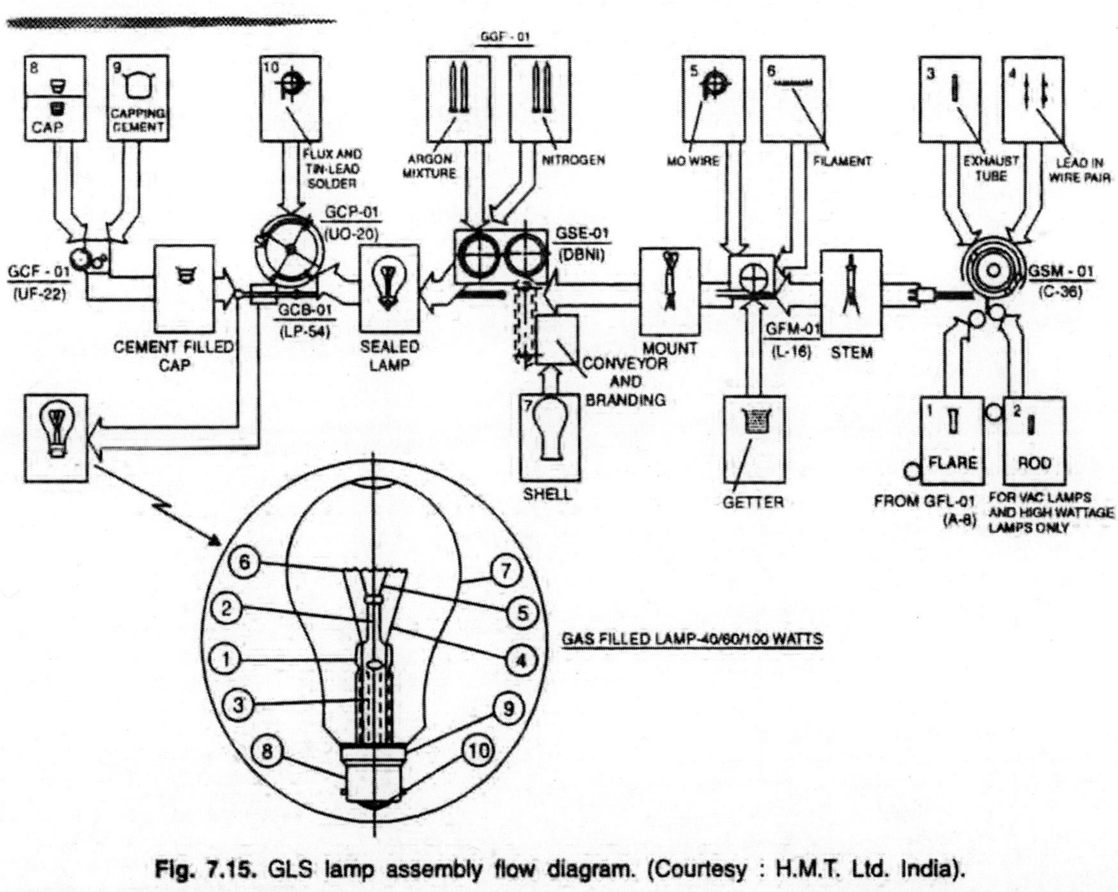

Fig. 7.15. GLS lamp assembly flow diagram. (Courtesy : H.M.T. Ltd. India).

number of operations. The early lamp making process was mostly manual and each maker had his own secrets of manufacturing process. Now International Electrotechnical Commission (IEC) has stipulated universal standards in regards to dimensions, lumen output, lumen maintenance, cap temperatures, cap dimensions etc. for control of interchangeability and safety. All these instructions have been included in Indian Standards and the manufacturing methods had to be adapted to incorporate all these standards which have also brought more similarities in the process of production.

Now automatic equipments have substituted nearly 90% of the manual operations. The internal assemblies of the components are carried out by automatic transfer chains and sealing, exhausting and capping operations are carried out on rotary machines with progressive operations at every station. For GLS lamps now-a-days chains are now available to produce even 4000 lamps in an hour with only 8 workers manning the principal stations. For mini and auto lamps even the glass blowing of the shells are carried out on fully automatic transfer chains where glass tubes are fed at one end and the desired shells are obtained at the other end.

Manufacturing methods are categorised in the following manner.

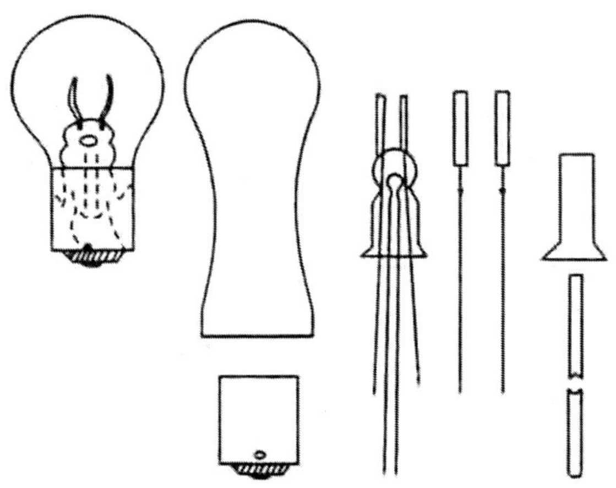

Fig. 7.16. Drop seal lamp components.

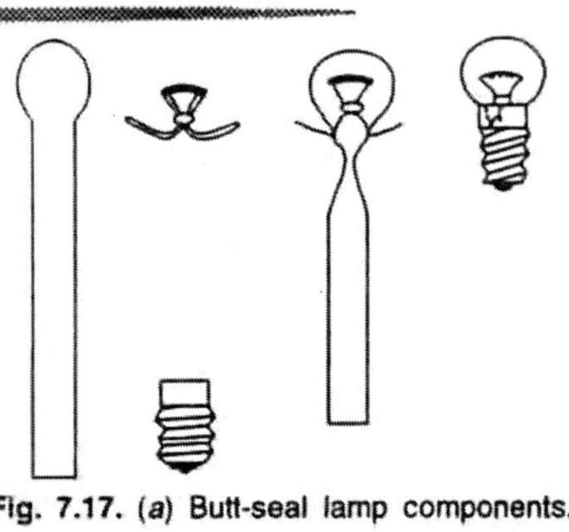

Fig. 7.17. (*a*) Butt-seal lamp components.

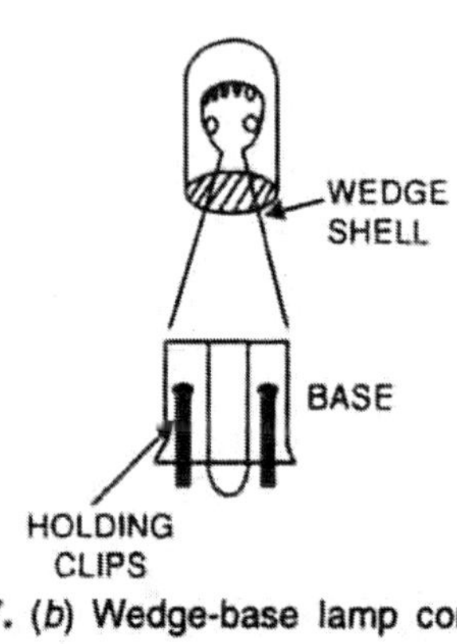

Fig. 7.17. (*b*) Wedge-base lamp components.

Manufacturing Methods

Drop Seal Construction

Sealing takes place between the flare and the neck of the glass shell. The outer skin of the glass shell after progressive heating drops on to the flare and the fusion takes place. This is applied for making Auto and GLS lamps.

Butt Seal Construction

A tube of approximately same diameter of the shell neck is sealed with the bcad mounted inserting placed inside the upturned shell. The tubing is constructed above the area of the seal for tipping off. This method is applied for making Mini lamps.

Wedge Base Construction

This is a further development on Butt Seal Construction. These lamps are cap-less types as used in indicator, panel and telephone lamps. Instead of shaping the lamp base to accommodate the cap. The neck of the lamp is heated and shaped as base for special holders.

See Fig. 7.16 and 7.17.

The lamp machineries are manufactured keeping in view the methods of operations those will be necessary to follow. Whatever methods might be followed, there are many operations which are common to all types of lamp making. The lamps can be made in fully automatic plants, semi-automatic machines or manually which depend on the availability of financial and labour resources.

Production Flow Chart

This flow chart covers almost all the operations needed to manufacture lamp of any category. The sequence of operations might vary as per mode of manufacture *i.e.*, manual.

Production Flow Layout

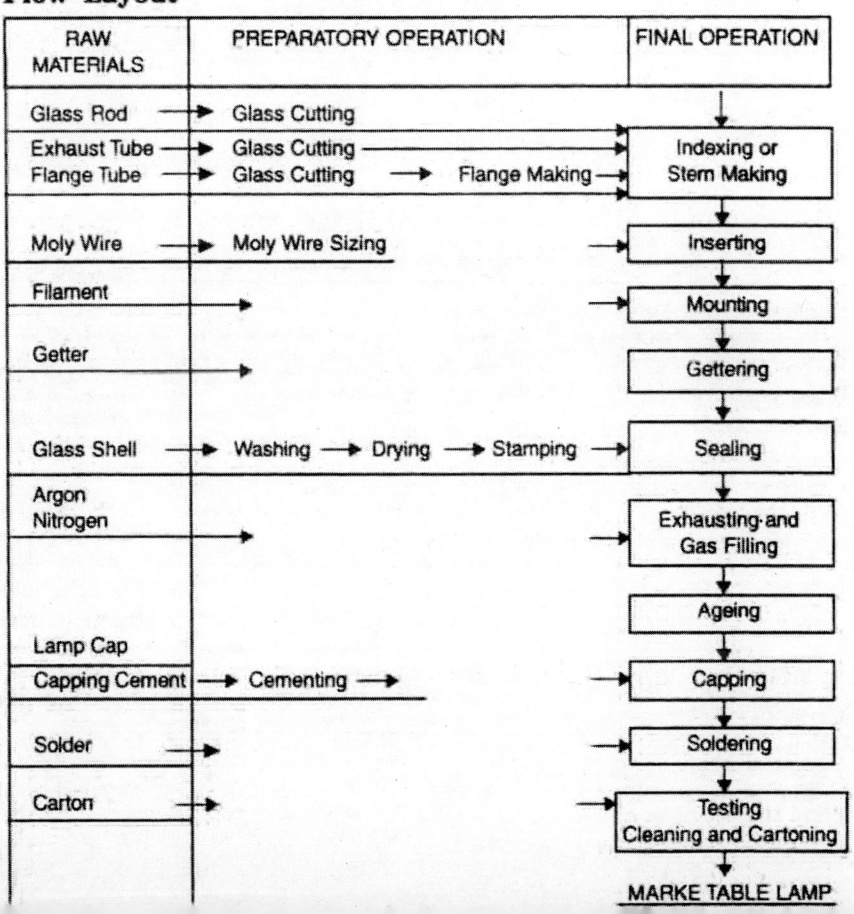

semi-automatic or fully automatic. The same is applicable for the manufacture of Halogen lamps also. Some operations as gettering, ageing, capping etc. are not necessary for certain types of lamps which are discussed in details later in this chapter. The manufacturer is to follow and alter this flow sheet to suit his product.

The following are the basic requirements of a lamp factory.

1. Dry compressed air in high volume under low pressure for slow moving semi-automatic machines and at high pressure for fully automatic service suitable for the efficient working of the burners.
2. High calory carbon gas under low and safe pressure.
3. Oxygen gas, argon and nitrogen gas in desired form.
4. Un-interrupted supply of electricity.
5. Clean water.

Raw Materials or Materials from End Product	*Mode of Operation Machine or Manual*	*Operation Number and End Product (Italic words)*
(*a*) Glass shell	Washing machine Drying machine	1. Acid, clean water washing, drying in the atmosphere, sun or by machine. *Washed shell.*
(*a*) Washed shell	Shell stamping apparatus or manual	2. *Stamped shell.*
(*a*) Full length (*b*) Glass rod (*c*) Glass tube (*d*) Exhaust tube	Glass cutting machine	3. (*a*) Cut to size. (*b*) *Sized glass rod* (*c*) *Sized glass tube* (*d*) *Sized exhaust tube*
(*a*) Molybdenum wire	Scissors and zig made of strong pins	4. *Sized molybdenum pig tails.*
(*a*) Sized glass rod tube	Flare machine	5. *Flared tube*
(*a*) Flared tube (*b*) Sized glass rod (*c*) Lead in wires (*d*) Sized exhaust tube	Indexing (stem) machine Annealing apparatus	6. Stem making *Annealed stem*
(*a*) Annealed stem	Button machine or manual with button die	7. Button making *Buttoned stem*
(*a*) Buttoned stem (*b*) Molybdenum pig tails	Inserting machine or manually with burner	8. Pig tails inserting. *Inserted stem.*
(*a*) Inserted stem (*b*) Tungsten filament	Mounting apparatus	9. Filament mounting *Flowered stem*
(*a*) Flowered stem (*b*) Gettering chemicals	Manual	10. Gettering *Gettered stem*

(*a*) Gettered stem (*b*) Stamped shell	Sealing machine	11. *Sealed lamp with extended exhaust tube.*
(*a*) Sealed lamp with extended exhaust tube.	Exhausting machine High frequency tester Argon-nitrogen Gas purifying circuit	12. Exhausting, Gas filling and tipping off *Gap-less lamp*
(*a*) Cap-less lamp	Ageing rack with volt meter and ammeter	13. Ageing the lamp *Aged lamp.*
(*a*) Aged lamp (*b*) Lamp cap (*c*) Cement	Cementing cap manually Capping machine	14. Lamp capping *Capped lamp*
(*a*) Capped lamp (*b*) Solder	Scissors Soldering apparatus	15. Cap soldering *Soldering lamp.*
(*a*) Soldered lamp	Testing holder	16. Final testing *Tested lamp.*
(*a*) Tested lamp	Manual	17. Cleaning and packing.

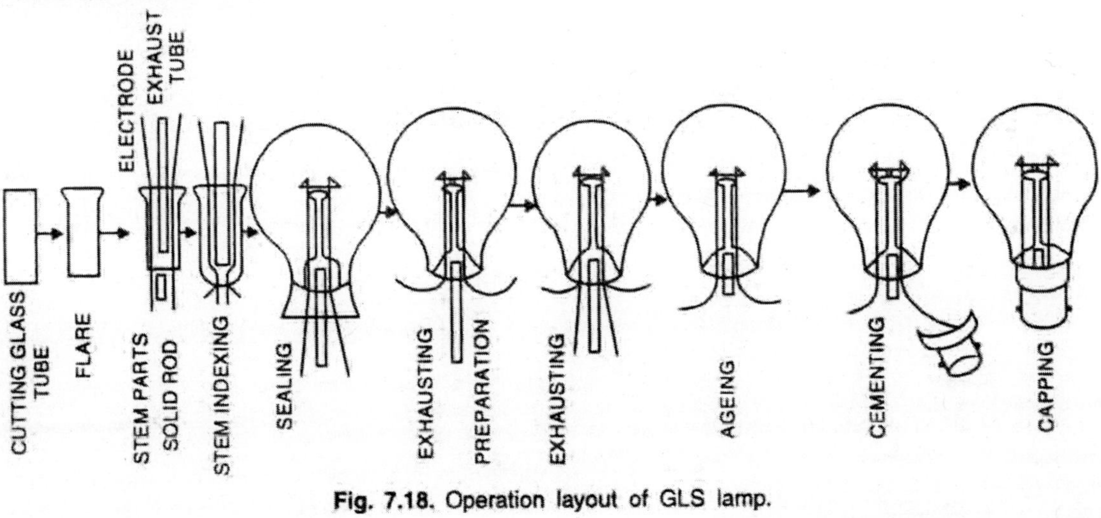

Fig. 7.18. Operation layout of GLS lamp.

SEQUENCE OF OPERATIONS FOR MANUFACTURING ELECTRIC LAMPS

The operations are described in detail as serially mentioned in the flow chart. Besides every operation the abbreviated form of each category of lamp is mentioned indicating the necessity for that operation.

Category	Abbreviation
G.L.S.	G
Auto lamps	A
Miniature lamps	M

The operations carried out by automatic machines are described separately. The description of the work involved at every station may vary from manufacturers to manufacturers of these machine but basically they do the same work. So, the knowledge of one set of automatic machine would provide sufficient guidance of the operating techniques.

Operation 1

Shell Washing and Drying (G.A.). The glass shells are dipped in water mixed with dilute hydrochloric acid or very rarely with dilute nitric acid, rubbed internally with brushes and then cleaned 2–3 times with clean fresh water. Special Shell Racks are made to keep the shell hanging from it with the neck pointing downwards. Wet shells are dried in the sun or by slow circulation of hot air. In any case, the drying should be a very slow process in a clean, soot or dust free atmosphere so that the evaporation should not leave any 'water mark' on the shell. If there be traces of acid inside the improperly washed shell, the bulb will not give proper light and it also would fail prematuredly. The shells of the 'Festoon Lamps' are not washed. Technical details of a 60 mm glass shell is given in Fig. 7.19.

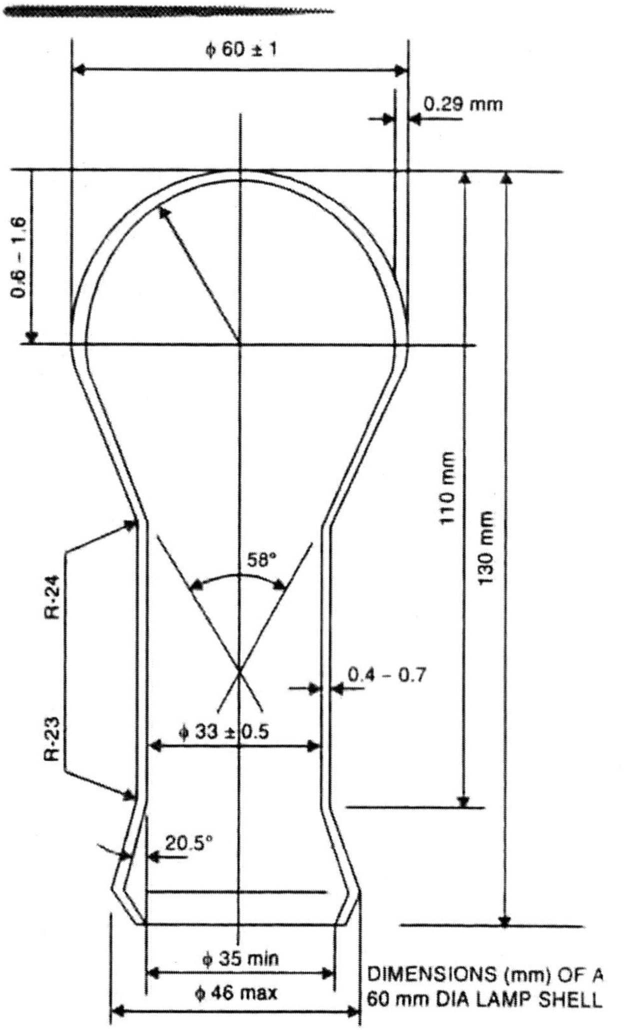

Fig. 7.19. Dimensions (mm) of 60 mm lamp shell.

Operation 2

Stamping on Shell (G.A.M.). At this stage the shell is first inspected of its cleanliness, snicks, cuts and blow bubbles. Branding is to be made distinctly and indeliably on to the lamp. Under no circumstances a lamp is to leave the factory without the prescribed markings. This operation is carried as per the suitability and arrangements of the other machineries. For the markings made on the outer surface of the lamp, the printing ink used is generally of ceramic base and gets itself embeded on to the lamp surface under heat.

Markings for GLS lamps :

The mark of the manufacturer.

The rated volts (marked as *V* or Volts)

The rated watts (marked as *W* or Watts)

Coiled coil or single coil

Gas filled (marked gas) or *V A C* for vacuum

Country of manufacture.

Markings for Automobile Lamps :

(The markings are made either on shell or cap.)

The manufacturer's name

Rated voltage (V)

Rated wattage (W)

Country of origin.

Markings for Miniature Lamps :

The minilamps which are made for automobiles as mentioned under its trade categorisation—Mini lamps from Group I to IIT are to be marked as the above 'markings for automobiles'. Due to shortage of space, only the rated voltage and wattage are stamped on the caps of the torch lamps.

If stamping is made inside the shell, this should be done very carefully leaving least possible printing material inside the shell, in fact this practice should not be encouraged at all. In the long run even a small trace of foreign material as left by printing material would seriously affect the life of the lamp.

Shell Unscrambling Washing, Drying, Stamping

(By Fully Automatic Machines)

The unscrambler unit is a conveyor of varying sections. At the loading position opened cartons containing fresh shell are placed which are picked up one by one by the transfer arms and emptied on to a belt conveyor. After going through various movements, the shells reach the gate mechanism which allows only one shell at a time to be fed into the holders of the bulb washing conveyor.

The bulb washing machine is a chain type indexing conveyor driven by seal/exhaust machine and fitted with carriers to carry the completely washed, dries and stamped shell to the loading station of the seal/exhaust machine. (*See Photo 7.3*). On the Bulb Washing Conveyor (BWC), the inside of the shells are washed with timed jets of water (two cold and three hot) and dried by passing over jets of hot air. These are stamped with statutory and manufacturers' markings prior to hot air operations.

Services required : Fuel gas—0.14 kg/cm^2 or 2 lb/sq. in.

Low pressure compressed air at 0.28 kg/cm^2 or 4 lb/sq. in.

High pressure compressed air at 5.5 kg/cm^2 or 80 lb/sq. in. and clean fresh water.

Full length glass tubes and rods are cut to size by glass cutting machine. The machine should be properly grouted and should be made vibration free. The carborandum wheel should not have any side play and should be kept in proper dressed condition with a diamond point dresser. Improper wheel and vibration cause small and minute cracks on the cut-ends of the glass which give away at the very first meeting with the heat from burner.

Use of Compressed Air and Fuel Gas. Till now the above operations did not have much use for 'compressed air' and 'fuel gas'. For melting glass, extra heat is necessary at the flame point of the burner. Compressed air is necessary to extract the full calorific value of the propane gas. In lamp industry, the glass being melted, needs gradual heating process where the volume of air supplied is more important than the pressure of the compressed air. It should be supplied through a reservoir to avoid the pumping impulses of the compressor unit. In the rainy season, the reservoir is to be drained after every 3 to 4 hours of operation to avoid condensation of moistures in the service lines.

Use of Fuel Gas. Every large factory has its own gas plant. In large cities there are arrangements for central gas supply. In India more factories run on indane gas cylinders supplied for domestic use. Whatever might be the source of supply, proper care must be taken of the fire hazards. Instead of connecting cylinders to individual machines it is advisable to install gas service pipe line and the cylinders connected at the starting point.

Due to insufficient supply of air, failure of electricity (compressor), there are violent blow backs and chances of heavy explosion. So, with every gas connection to the machine, a 'flame arrester' must be used. Many patented and sophisticated flame arresters are available in the market. A simple one can be locally made as shown in the sketch below.

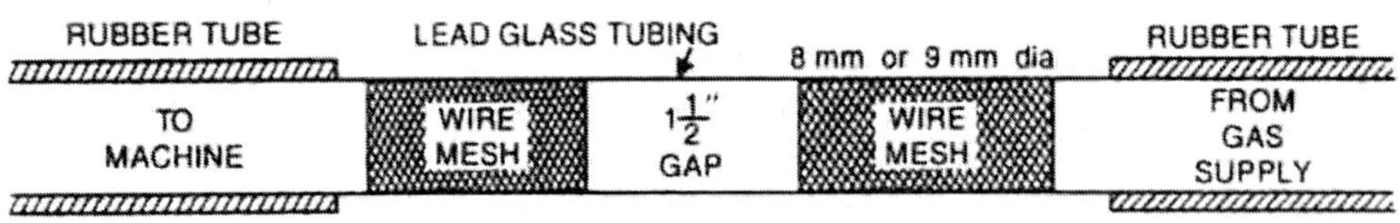

Fig. 7.20. Indigenously made flame arrestor.

Operation 3

Glass Blowing Operation (M). For miniature lamps as festoon lamps and torch lamps etc. the shells are made by blowing 3 mm to 4 mm dia lead glass tubings. One end of the tube is progressively melted over the flame and the other end is either connected to the compressed air line for blowing or mouth blown by an operator. Here proper dies should be used to get uniform shape of the shell. After shaping the blown end the shell is cut off over the neck area. In mouth blown shells the loss percentage is very high. It is advisable to use compressed air for this purpose where the tube is placed in a rotating chuck, blowing carried out through a semi-rotary valve for control of compressed air supply and shaped by means of a pair of spring loaded dies.

Fully automatic machines are available in which sized tubes are fed at one end and perfectly shaped, sized mini shells are received at the other end.

The thickness of the mini shell walls are very thin. Specially in mouth blown shells the wall remains very uneven. These shells expells heat very quickly and in the process produce fine cracks. The mouth blown shells need annealing for longer period after blowing which is generally neglected by the operator. The blowing stress must be removed properly and the hot shell should be placed inside an asbestos-lined container allowing it to cool off. gradually to ambient temperature.

Operation 4

Cutting Molybdenum Wires to Size (G). Molybdenum wires are used to support the filament and prevent the filament from sagging. It also gives rigidity to the otherwise very brittle condition of the filament after ageing. In any case it is to be understood that it carries out the heat from the filament and reduces the lighting efficiency of the lamp. The minimum number of these supports used with minimum length-increase the life of the lamp in the long run. For the sake of beauty it is seen that a good length of the tail is kept at the rounded end which actually acts as an efficient heat expander inside the lamp and helps in reducing walkness of the filament contact point inside the rounded end.

After cutting to size, the support wires with a rounded end look as 'pig tails', hence it is nicknamed as such. The diameter of the moly wire, the length and the numbers of them to be used all depend on the wattage of the lamp, the length of the stem and the geometric centre of the lamp.

Operation 5

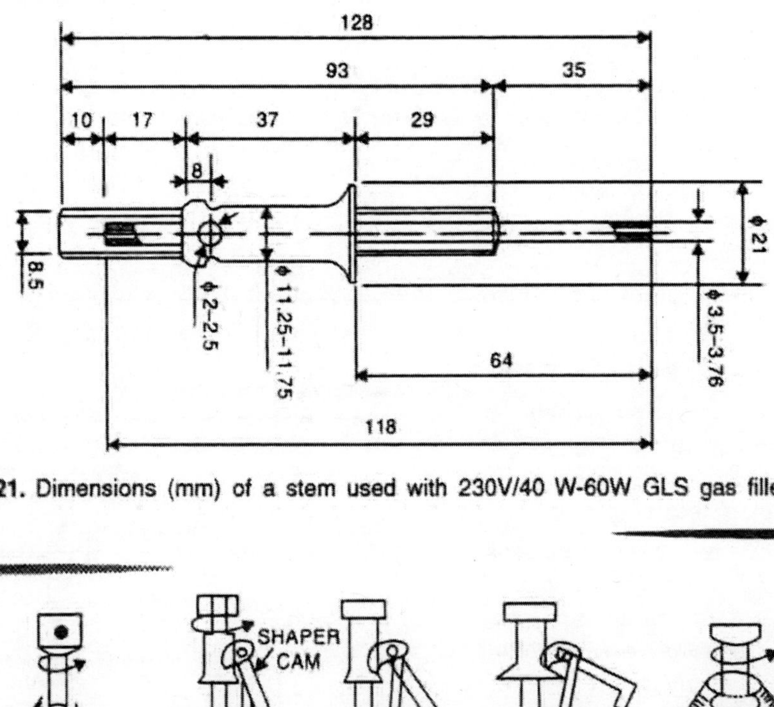

Fig. 7.21. Dimensions (mm) of a stem used with 230V/40 W-60W GLS gas filled lamp.

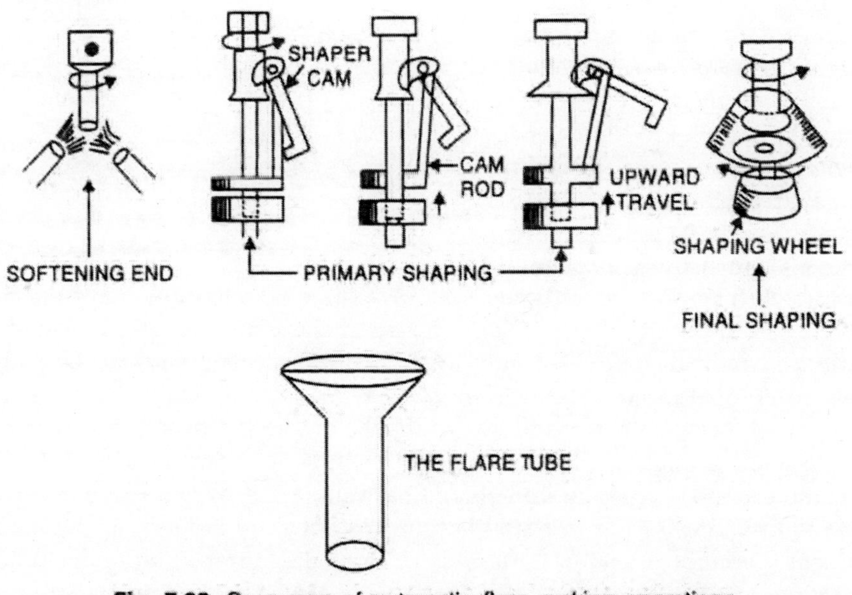

Fig. 7.22. Sequence of automatic flare making operations.

Flaring (G.A.). Flange is the base of a stem on which all the internal components of a lamp are built up. When one end of a lead glass tubing is heated and flared outward like a 'funnel', it is called a flange. The degree of the angle upto which it is to be bent is of utmost importance. While sealing, the outer periphery of the flange only, is met by the heat-softened glass neck of the glass shell. If the flange angle is too large, say around 90 degrees with the central axis of the flange tube, then the flange periphery meets the glass shell too early. This causes bad sealing leaks, bad seats for the caps and sometimes flange-cracks. On the other hand if the angle is too less, the periphery meets the shell too late which cannot catch up with the already running down melted ring piece of the shell. Good sealing can never take place in such condition. The ideal angle lies between 45 degrees to 60 degrees which is to be adjusted with every type of shell to be used. (*See Photo 7.4* and Figs. 7.21 and 7.22)

Operation 6

Indexing (Stem Making) and Annealing (G.A.). Literary meaning of the stem is the 'main body of a plant springing from a root'. The stem is the main body of a lamp consisting of the flange, exhaust tube, solid rod, and lead-in-wires (Fig. 7.23). A stem machine or an indexing machine assembles all these items together. A stem has to perform the following functions :

1. To form a leak proof joint between the atmosphere inside and outside the lamp.
2. To provide means to evacuate the inner atmosphere of the lamp and also to provide passage for the introduction of inert gas.
3. To provide insulated holder for lead-in-wires at the pressings, support for the pig tails and also provide passage for the electrodes up to the lamp cap.

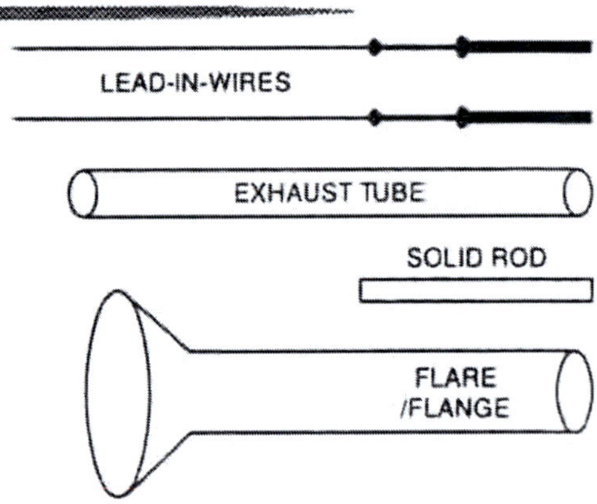

Fig. 7.23. Exploded view of parts of a stem/indexing.

For executing all these functions, the stem becomes the most important 'assembly of components' in a lamp. Fully automatic stem machine assemble all these components at different stations where components in bulk are fed by one worker. Generally the stems for higher wattage lamps are made manually in single-head power operated machines. In semi-automatic machines, the components are to be fed by 2 to 3 workers at different stations. The stem operator plays the most vital role for producing good quality stems. Proper fusion of the plain end of the flange with the exhaust tubing and rod can be judged only by a very experienced worker. The adjustment of burners at every operational station for gradual melting of glass is to be done very carefully. Once these settings are made, these should not be disturbed much. Through the pressings, the 'dumet' portion of the lead-in-wires pass; so its 'spot welded point' should be sufficiently inside the stem pressing and should be well sealed by the fused glass. The first and foremost requirement for the pressing is to make a leak-proof joint of the lead-in-wire. All the slow leaks occur (say after 20% of lamp life) due to the faulty installation and adjustment of dumet in the pressing. If the dumet weld spot is too much inside the pressing, a great amount of heat of the filament will be carried inside the pressing and the rounded spot will try to bore a hole inside the glass causing slow leak. The adjustment for this differs from wattage to wattage and can be obtained through few trials. (*See Colour Photo 7.5*)

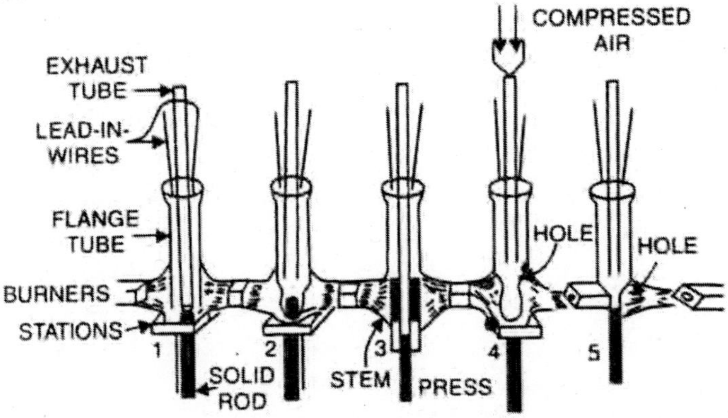

Fig. 7.24. Sequence of operations of Indexing/stem making.

The stem machine burners are fed with gas, oxygen and compressed air. All these three items are to be adjusted to get the proper flame and force keeping in view of the following discrepancies.

(*a*) If less oxygen and more air is used, the stem pressing will be black in colour—the melting takes more time and the shape will be erratic. This causes slow leakage at later stage.

(*b*) If more oxygen and less air is used, the pressing will look milky white, broken white or crystalline structures inside. This again is a bad stem which is too much hardened due to excessive heat of extra oxygen from burners. This type of stem cannot be annealed properly it gets cracked while sealing. At the last operation station, pre-heated compressed air should be fed for bursting hole, shaping of the stem neck, if cold air is used, this again results in stress formation which ultimately causes the stem to crack.

The melting temperature of glass is 1400°C. In stem making this temperature is reached for a very short period. The heated portion of the flange tube, exhaust tube and solid rod all fuse and melt together when the 'pressing arm', press them all to form into a single unit namely 'the stem press' or 'stem pinch' (See Fig. 7.24).

After the pressing and bursting operations, the stem goes to the next station of operation still in red hot condition. The stem has got to be lifted out from the machine to leave the stem station free for next refilling. So, and 'annealer' is placed near to this station where the stem is lifted out and placed on to the annealer rails. In the annealer the stem slowly moves through gradually decreasing heat zones and finally to the room temperature. It is advisable to put the hot stems inside a box atleast for two hours before inspection. The annealer heat is to be adjusted properly. In linear transfer automatic machines when the stems pass on chains through annealer oven the temperatures always are to be kept under strict control. On sudden chilling the stems will become very brittle and it might give away in any stage of production and it then become very difficult to trace the defect due to improper annealing. The chances of hard stem is more on semi-automatic radial machines and the annealer has go to be placed nearest to the stem machine.

Operation 7

Button Making (G). Practically the molybdenum support wires provided to support the long filament in ship-shape condition act only as deterrent to the efficiency of an electric lamp.

In fully automatic machines, the end of the solid rod is heated by the burner and after a predetermined period a two piece die forms it to a base for inserting the end of the pig tails. Soon after the die is removed the moly wires position themselves automatically and their hot ends are pressed into the hot button.

For manual operations, the solid rod end is heated on the flame by a worker till it reaches red hot condition and then it is pressed on to a die to form the button. If the solid rod is brought too near the flame, the button will be black in colour which is undesirable. The solid rod end should be heated on the tip of the flame to get a fine colourless button.

For fully automatic machines the operations 7 and 8 are carried simultaneously. For manual working, button making and inserting are different operations but may be carried out by the same worker.

Operation 8

Inserting (G). (*See Colour Photo 7.6*). Inserting by automatic machine is described under button making. Generally after making the button, the same worker starts inserting the pig tails around the periphery of the button. This is a very simple operation. If a pig tail is put on to a under-heated button, this gives away in the running life of good lamp and hangs down with the filament causing it to fail early. Every pig tail must be given a little pull to test its strength after inserting. Also the burnt out pig tails due to over heating are again injurious to lamp life.

Operation 9

Mounting (G.A.M.). This is the process of installing the filament between the ends of the lead-in-wire (electrodes).

In a fully automatic machine, the operations are carried out as seen in the figure. The transfer chain carries the completed stem to the mounting station *'A'*. At *'B'*, the electrode end is flattened and the filament tails are pinched on to them. Then the electrodes with the filament are bent upward *'C'*, the solid rod is heated and a button made with a die (these operations are described earlier to make the readers conversant with those operations). At *'D'* the moly pig tails are inserted in predetermined numbers and cut to the desired length, at *'E'* the filament is brought to the final position and the moly pig tails are looped around the filament.

For mounting manually, the worker at first flattens the ends of the electrodes with a foot operated mounting apparatus. He then encloses the unlooped ends of the filament by bending the flat ends of electrodes using a little pressure either with the mounting press or by a pair of nose-end pliers. Here the pressure for holding the filament is a critical action. The filament is then passed through the moly pig tails and the pig tails are shaped as needed.

(*See Photos 7.7, 7.8 and 7.9*)

Mounting for auto lamps is carried out by spot-welding the filament ends with electrodes.

For miniature lamp, the stem pressing is replaced by a bead of sintered glass as the anchor for the dumet wires which also act as an insulated separator between the dumets.

Photo 7.3. Automatic bulb wash conveyor and bulb stamping unit.
(Courtesy : Badalex—A Div. of Sale Tinley Technology, England.)

Photo 7.4. Horizontal flare machine.
(Courtesy : Badalex—A Div. of Sale Tinley Technology, England.)

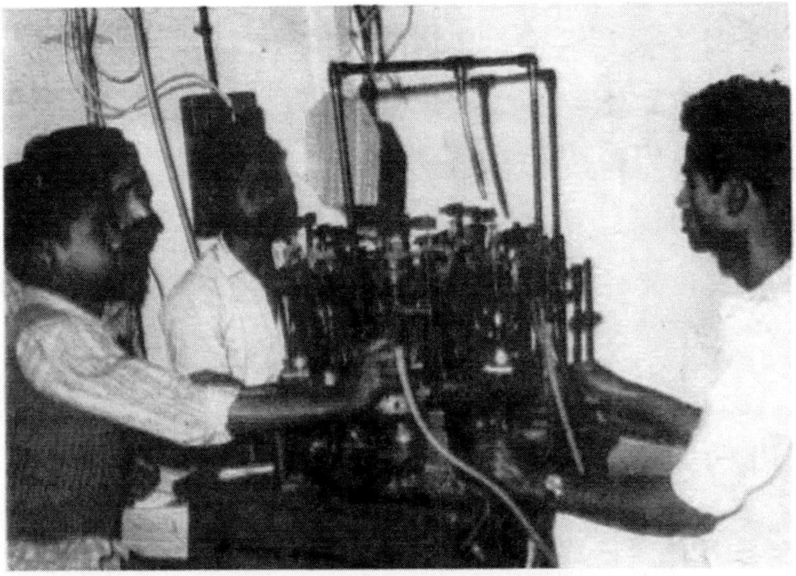

Photo 7.5. Stem making with a semi-automatic machine.
(Courtesy : Sudip Electrical Works, Calcutta.)

Photo 7.6. Inserting and mounting in a small scale factory.
(Courtesy : Navjyoti Lamp Works, Patna.)

Photo 7.7. 32 station automatic stem machine.
(Courtesy : Hindustan Machine Tools Ltd., India.)

Photo 7.8. 24 station filament mounting machine.
(Courtesy : Hindustan Machine Tools Ltd., India.)

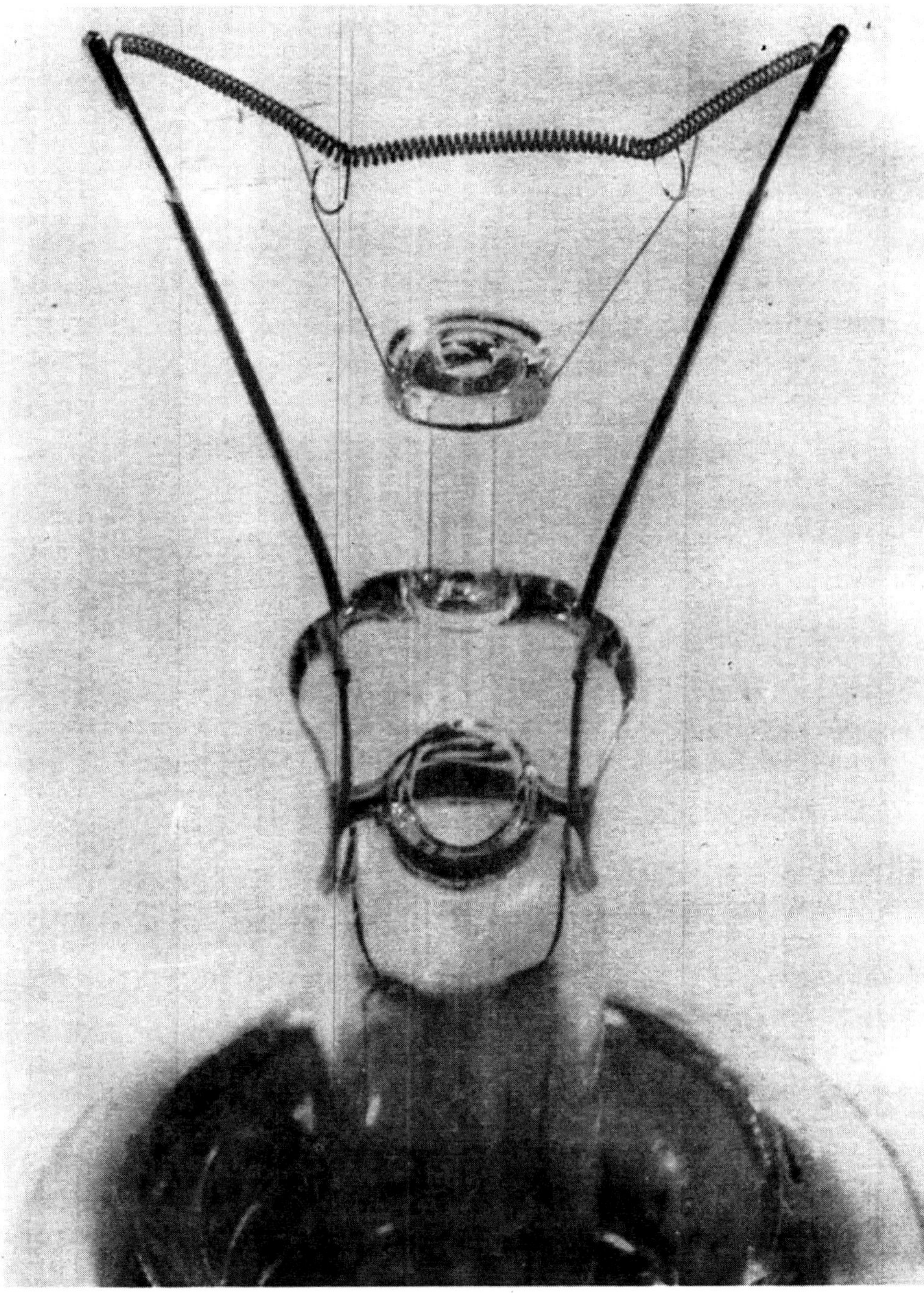

Photo 7.9. A perfectly mounted filament with moly-wire supports. (Courtesy : Philips Tech. Review, Holland.)

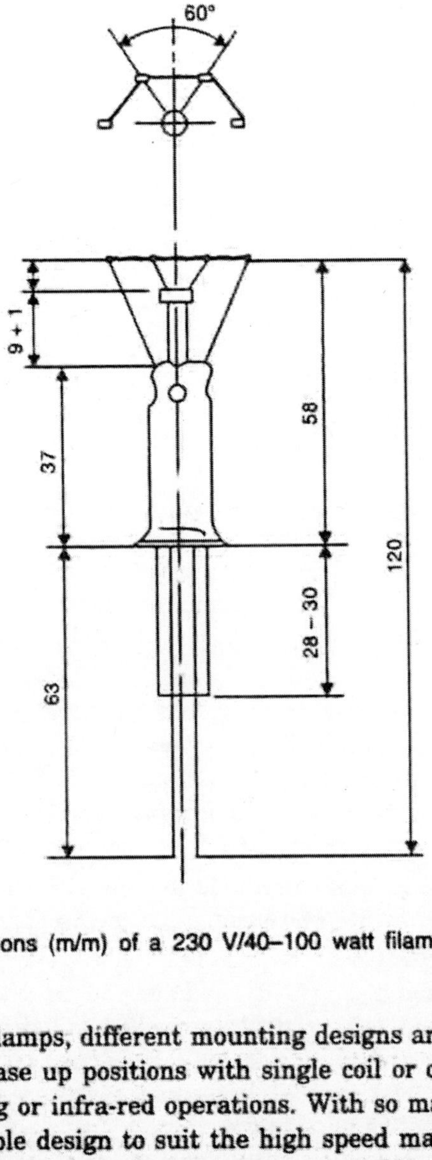

Fig. 7.25. Dimensions (m/m) of a 230 V/40–100 watt filament mounted stem.

For different types of lamps, different mounting designs are employed. The lamps may be used with base down, base up positions with single coil or coiled coil filaments to serve in vibrational, flood lighting or infra-red operations. With so many variables, a compromise has to be sought for a simple design to suit the high speed machineries. The mounting for higher wattage lamps need special care as single coil filament are used for these lamps. Lamps from 300 watt and above, the button and the solid rod must be made with pyrex or corning glass. Also a mica disc is placed at the neck of the lamp, so that more heat may not reach the cap to overheat it.

After completion of the filament mounting, the assembly is called 'the flower.'

Where mounting is done manually, it should be carried out in a dust free, well lighted room with skin-gloved hands.

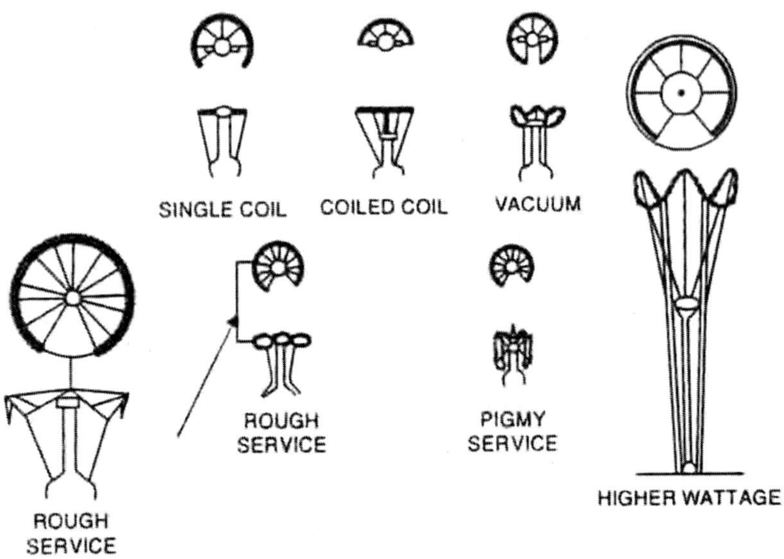

Fig. 7.26. Different types of filament mounting.

Operation 9(*a*)

Auto Stem and Mount (Automount)—Fully Automatic Machines. The *Photo 7.10* is a typical example of an ultra high speed 73 station linear indexing machine which combines the operations of stem making, filament mounting, lead-in-wire manipulations, and filament mounting on one machine. Even the loading of the components and finally unloading of the mounted stem are also automatic. Here the first chain with 73 heads is placed above the second chain with 49 heads. The second indexes between the smaller centres receiving timely drive from the upper chain when it is to be actuated.

The 49 heads of the bottom chain hold lead-in-wires in die blocks during stem making upto 48th station till annealing and completion of the stem making operations when it is separated from the upper chain.

After the chains have disengaged, the heads on top chain hold the stem in position so that the processes of lead wire spreading, cropping, flattening, hooking, button forming, filament feeding and filament mounting could be carried out. After gettering it unloads into turnover turret. Various detection and inhibiting devices are incorporated to sense misfeeding and malworking of the stations due to which fresh loading and working on defective stations will not be carried out, thus eliminating wastage of materials.

Utilities needed. Fuel gas at 0.21 kg cm^2 (3 lb/sq. in.).

L.P. Air at 0.28 kg cm^2 (4 lb/sq. in.). Oxygen at 0.35 kg cm^2 (5 lb/sq. in.).

H P. Air at 5.5 kg cm^2 (80 lb/sq. in.) and 3 phase electricity supply.

Station Layout of the Automount Machine :

1. Blank
2. Flare feed (jaws open)
3. Blank
4. Flare detect
5. Wire feed
6. Wire feed

7. Electrode detect (cooling starts)
8. Exhaust tube feed (jaws open)
9. Exhaust tube detect
10–12. Fire Ex. tube push down
13–46. Fire (flare setting)
14–16. Ex. tube glaze
19. First pinch
31. Second pinch
32–35. Blow hole and flare (hot air) and third pinch
41–46. Annealing
47 and 48. Blank
49. Electrode flatten
50. Electrode hook
51. Auto filament feed clamp (cooling ends)
52–54. Blank
55. Electrode twist
56. Electrode push back
57–61. Button forming fires
62. Button with support wire insert
63, 65, 68, 72. Blank
64. Support wire hook and set
66. Set electrode forward
67. Support wire curling
69. Final electrode setting
70. Getter
71. Getter drying
73. Unload into turn over turret of seal/exhaust machine.

Operation 10

Gettering (G.A.M.). Even after taking all the precautions of cleanliness and removal of impurities, the enclosed area with all the components inside a lamp is liable to be contaminated with many foreign matters, specially at high temperatures, the filament and the moly wire supports may leave off small amount of oxidized skin, the inner surface of the glass shell might have some residual oxygen molecules or the inner atmosphere itself may not get fully evacuated of some oxygen. To neutralise all these impurities and hold them in non-reactive suspension some chemically very reactive substances are introduced inside the lamp, known as 'GETTER.' In many cases some pure metals are also used as getter to absorb free oxygen inside the lamp. For auto lamps zirconium is an widely used metallic getter. Barium and magnesium getters absorb the impure gases and get themselves collected on to the neck of the lamps. For GLS lamps finely ground red phosphorus has been found to be very suitable. Red phosphorus is used with a mixture of cryolite and chloroform spirit as its carrier. For auto lamps fully powdered zirconium with collodion and amylacetate make very suitable getter. Generally gettering is not necessary for very low watt vacuum mini lamps where chances of residual oxygen is very less due to its small dimension and high vacuum.

Gettering is again another very important but generally neglected operation in lamp making. In most of the factories it becomes a normal habitual function for which continuous vigilance is thought to be unnecessary. It is to be remembered that the life of a lamp depends on getter, the quality of the light emitted by the lamp also is dependent on good gettering.

There is no particularly fixed getter which can be written down as a recipe. In India where we have very humid and dusty atmosphere the getter is always changed to suit the occasion. It is advisable to use a chemical getter in combination with a metallic getter. The amount of getter required in a particular type of lamp again cannot be fixed as this also would vary from month to month depending on weather. Gettering is not all that good as thought by many lamp makers. Excessive getter is a sure lamp killer. As silvery or red shade, the excess getter not only reduces the quantity of light but acts as impurity and starts shredding of the filament skins thereby reducing the lamp life. With little perseverance and methodical trials, the quantity of getter can be easily found.

Inside a vacuum lamp, there is no gas and the chances of residual oxygen is lesser than a gas-filled lamp. A vacuum lamp needs a lighter getter. Also the single coils being longer in length the filament catches more getter in a vacuum lamp. For this reason a vacuum lamp should be soft gettered.

In a gas-filled lamp, the inert gas might carry some oxygen with it. So, it needs hard gettering with assistance from crayolite. Hard gettering does not mean that very thick gettering mixture is to be used. It should be a little thicker than the vacuum getter. For gas filled lamps, a simple touch of the corners of the filament into the getter mixture gives the desired result—if the coiled coil filament is dipped fully into the mixture, quite a good amount of getter enters the coil resulting a red lamp.

As carrier agent for getter, amylacetate, chloroform spirit and sometimes liquor ammonia is used. The carrier only deposits the getter on the lamp components and it completely vaporises during sealing and exhausting operations. So, getter comes into action before the lamp is first lighted during ageing action or at the initial sparking in the exhaust bench.

A lamp is always to be aged with the base up position, so that the getter vapour can get deposited in the neck and inside the cap of the lamp. As the gettering in the vacuum lamp is thin and also there is no inert gas present inside, the voltage is to be stepped up slowly which will cause slow burning of the getter material and the fumes from it will rise slowly towards the lamp cap. In gas filled lamps, the ageing is to be carried out quickly before heating up of the inner atmosphere which will then carry the getter fumes all over the lamp. These operations are carried out under ageing but the gettering worker should know about the after effects of his action hence these are described here.

After gettering the filaments, the flowers are to be sealed on the same day and the sealed lamp also is to be evacuated soon after sealing. It has been seen that sealed but unevacuated lamps after 48 hours had a very high percentage of failure at the ageing stage.

Where the diameter of the filament wire is thick, getter is applied on the side of the electrodes with a painting brush.

For auto head lamps this gettering acts as surface gettering throughout the whole life of the lamp.

Whenever trying a new batch of getter, it is advisable not to put it to production immediately. At least two to three dozens of lamps are to be made with it before final use in the production line.

In small factories red phosphorus is ground in a smaller, in solution with carrier chemicals. Fine grinding and proper consistency can never be obtained by this process. Instead of making the lamp free from foreign materials, it creates more impurities inside the lamp. A small stainless steel drum 12" length and 6" diameter can be used to make good getter. The drum is to be filled in with 3/4" stainless steel balls up to it's 3/4th capacity. Then proper quantity of red phosphorus, chloroform spirit, crayolite, a little barium etc. are charged into the drum. This tank is to be run on a $\frac{1}{2}$ H.P. Ball Mill Rollers for a minimum of 48 hours. The getter thus made will be of real good quality.

Gettering must be done by the same person to complete the same batch with which he had started working.

Operation 11

Sealing (G.A.M.). The process of covering the flower with the glass shell by joining the outer edge of the flange with the neck of the shell is known as sealing. At this point the assembly takes the shape of a lamp. In a rotating chuck, the flower is placed on the central head and the shell is placed on to a properly adjusted collar so as to rest the flower approximately at the geometrical centre of the shell-dome. The outer skin of the shell is gradually heated by sets of burners at different stations. The outer skin under influence of gradual increase of heat starts sagging and finally meet the outer edge of the flange of the flower. On further heating, the flange edge also melts along with the shell skin and forms a leak proof joint. The bottom portion of the shell drops off.

(See Photo 7.11)

Here the adjustment of the burners play the most important role. The speed of rotation and the change over time from station are fixed; so the burners are such adjusted that the operation at every station is completed just before the change over to the next station. At the last station, the shaping of the lamp takes place with pre-heated compressed air introduced through the head.

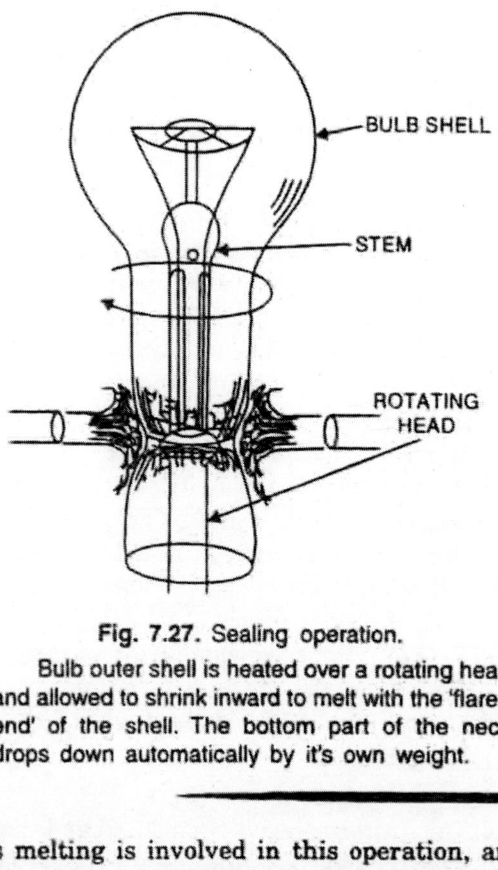

Fig. 7.27. Sealing operation.
Bulb outer shell is heated over a rotating head and allowed to shrink inward to melt with the 'flared end' of the shell. The bottom part of the neck drops down automatically by it's own weight.

Weather the sealing is carried out manually, or automatically, the above sealing functions remain the same for all. Because glass melting is involved in this operation, an experienced person is necessary to supervise these actions and adjust the flames daily.

Proper care should be taken in the rainy season to supply dry and moisture-free air to the burners. Odd sealing shapes will occur due to uneven and erratic flame from the burners caused by the presence of moisture in compressed air which go generally unnoticed. In the cutting burners, it is advisable to use oxygen in the semi-automatic machines which saves times and increases production. For manual centering of the stem, the shell should be given a circular motion and the adjustment carried out with exhaust tube. By this process the straightness of the filament and the centering can be observed from all the angles.

Shells without solid rings at their mouth are not suitable for machine operations as they need to be pinched by the operator for starting separation of the bottom portion. The coloured small shell glass is very hard and should be melted with air and oxygen. Uneven shell wall, improperly annealed glass shells cause many sealing problems. For GLS lamps, the shells should conform to the I.S. Specifications as in IS : 1112 : 1963.

In miniature and festoon lamps, a length of tubing of approximately the same diameter as the blown shell is joined with the bead mounted flower. The ends of the dumet are splayed out. The tubing is placed inside the blown shell and sealed over the flame. When the sealing is completed, the tube is tipped off above the constructed area to facilitate exhausting.

Sealing make the neck of the lamp very hardened. A sealed lamp must not be exposed to colder atmosphere. Fine cracks develop due to sudden cooling which increases the percentage of rejection in the exhaust bench.

In radial type of machines, the collars on to which the shells rest get heated soon and new shells get cracked as they are placed on them. The collars should have asbestos lining to overcome this difficulty.

Operation 12

Exhausting (*See Colour Photo 7.12*). In lamp making, next to importance in stem operation is 'exhausting' or evacuating of the lamp. In automatic radial machines every station is evacuated by individual vacuum pump which greatly reduces the operational time. In semi-automatic linear type benches one powerful vacuum pump evacuates one vacuum, manifold on which several heads are erected to connect the lamps for exhausting. The number of working heads depend on the capacity of the pump and the type of the lamps to be evacuated. The open end of the exhaust tube of the lamp is directly melted on the heads of the manifold or push-fitted on to the rubber tube of the manifold head. Whether the exhaust tube is melted or push-fitted, the main criteria is that it must form a completely leak-proof joint. A little negligence at this spot causes heavy loss of production and down time of the machine. Due to continuous use, the holes in the rubber head get enlarged or sometimes fissured with sharp glass particles which go undetected; so it needs frequent testing. In the directly melted joints, great care should be taken of the previously made two or three joints, which get very easily cracked and form minute leaks. Actually many exhaust bench workers think it to be a matter of pride for them and a 'show of skill' when they operate the manifold head with 6 to 8 old joints. This practice is not to be encouraged in lamp making.

A piezo electric coil is used for testing the colour of the atmosphere inside the lamp. This tester is heavily insulated and the end of the prod is only used for action. This produces very high voltage and always should be kept dry. The high voltage static electricity charges the atmosphere inside the lamp and produce the following colours :

1. If there be heavy leak inside the lamp meaning presence of atmosphere in full strength, the prod will produce no colour inside the shell.

2. If no leak is present and the vaccum pump works; a small touch of the tester on the top globe of the lamp will show violet colour inside the lamp. This violet colour indicates that the joints are leak-proof, the pump is working and there is air inside the lamp.

3. As more and more vacuum progresses, the violet colour will diminish and the glass shell will show some greenish stinge. At this stage pure nitrogen is introduced and the evacuation process continued which carry away the residual gas from inside the lamp (known as flushing).

4. When the lamp is rightly heated and properly evacuated, there will be no colour inside the lamp but the stem pinch will show a flickering yellowish glow meaning the pump has worked well and it is time to shear off the lamp by melting the exhaust tube near the base of the lamp.

5. Wherever necessary argon-nitrogen mixture is to be filled in when the lamp has reached the condition as mentioned in No. 4 because the colour of argon-nitrogen will show again violet stinge just as the presence of atmosphere inside the lamp. In a well filled lamp there will be fine strands of violet discharge from the moly wire ends and electrode ends to the lamp dome.

The discharge strength of the tester is to be adjusted as per the size of the lamp and also the place of it's use. A powerful tester is never to be used over thin places and never on to the head joints.

Correct heating of the lamp is one of the most important operation. In gas-filled lamps, the impurities and the residual gases are cleared off by giving the lamps a wash before filling or during heating but in vacuum lamps this advantage cannot be obtained. This necessiates more heating time for vacuum lamps. On the other hand if any lamp is over-heated during vacuum, they loose their normalising qualities and tend to be brittle with the first drop of rainfall on them. In any case the lamps need be properly heated on the exhaust bench otherwise the life of the lamp will be much reduced and they will get blackened very shortly. A properly heated lamp will show yellowish glow near stem as stated earlier.

Exhausting operation means creating vacuum inside the lamp. This is carried out by rotary oil pump. The serviceability of the pumps play an important role in making good lamps. Before going into details of pumps some useful units for calculation of vacuum conditions ought to be known.

The unit of vacuum is 'Torr'. It is named after its discoverer Torricelli of Italy. At normal temperature and pressure, if an one metre long tube with one end closed having one centimeter inner diameter is filled with mercury and inverted on to a trough filled with mercury, it would be found that the mercury inside the tube would stand to a height of 760 millimeters. The column of mercury inside the tube stays due to the air pressure on to the surface of mercury in the trough. This 760 millimeters of mercury has been named as 760 Torr. Now from the above fact it can be easily surmised that the column will come down if the air pressure on to the surface of the mercury is reduced. This reduction of pressure is none other than creation of vacuum. So, when we say 200 Torr, it would mean creation of such vacuum that the mercury column would come down to 200 millimeters from 760 millimeters.

Actually vacuum till 1 Torr is considered to be very low or inferior vacuum in the lamp industry. Fractional torr vacuum strength is used in the lamp industry. A vacuum of 1/1000 m/m of mercury is accepted as good standard. This is known as medium vacuum. The following chart shows vacuum characteristics and how a little failure in vacuum affects the lamps.

Characteristics	*Vacuum*			
	Low	*Medium*	*High*	*Ultra high*
Negative pressure-Degree of vacuum in m/m of mercury	760 to 1 mm	1 to 10^{-3} mm	10^{-3} to 10^{-7} mm	10^{-7} and less mm
Nos. of molecules present per cubic meter	10^{25} to 10^{22}	10^{22} to 10^{19}	10^{19} to 10^{13}	10^{13} and less

In lamp industry, the double stage rotary pump creating a vacuum of 1/1000 mm (10^{-3} Torr) of mercury column though seems to be good enough would still leave a large amount of oxygen molecules inside the lamp. From the above chart it can be calculated that a GLS lamp having 120 cc internal volume would still have 300,000,000,000,000 molecules of air left inside the lamp even after evacuated by a two stage pump. From this figure the importance of heating the lamp, gettering the filament and keeping of the pump in top serviceable condition could be understood. While burning, the tungsten filament gets oxidised and the oxygen present around the filament is responsible for it. The residual presence of oxygen are not detectable by the electrostatic tester. A lamp made without proper getter and heating and evacuation or gas filling might apparently seem to be perfect on immediate tests but in the long run it would get blackened prematuredly or fail suddenly. This is one of the worst fault in lamp making when proper instructions in the methods of operations are not being followed.

For accurate determination of the degree of vacuum and inert gas pressure inside the lamp, in most of the factories a combined-vacuum-pressure gauge is used. Actually a broad dial finely graduated vacuum gauge is necessary. The gas-filled lamp always has a little vacuum inside the lamp to provide space for expansion when in operation. This is generally filled with 700 mm and 720 mm of mercury. For having a combination of pressure gauge the range becomes wider and finer graduations on the vacuum side are being sacrificed. Practically a vacuum gauge in combination with a mercury column made and graduated locally for each individual machine gives best performance and would be the right answer. In the mercury column 720 Torr or 720 Torr can be easily marked and the operator will have no difficulty in filling the lamps till this level is reached.

THE VACUUM PUMP

A two stage rotary type vacuum pump unit is described here. The pump unit is immersed in oil inside the rectangular chamber *H*. Generally heavy duty rotella oil is used which maintain high viscosity even at high temperature. The oil also ensures the leak proofness of the assembled parts and acts as coolant at the same time. (Fig. 7.28)

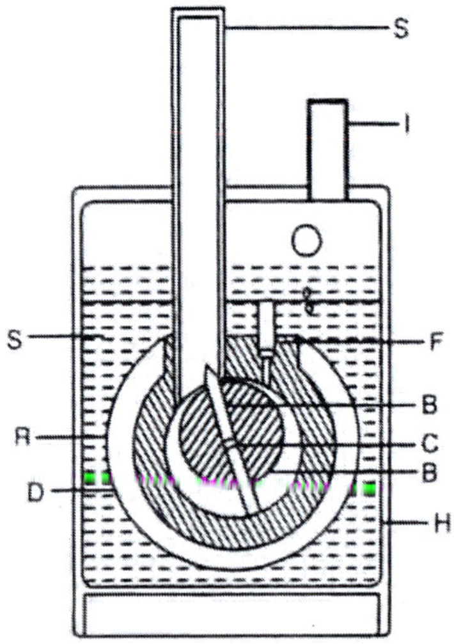

Fig. 7.28. Internal constructions of an oil immersed vacuum pump.

Photo 7.10. 73 station linear indexing machine with automatic filament feed unit. (Courtesy : Badalex—A Div. of Sale Tinely Tech., England.)

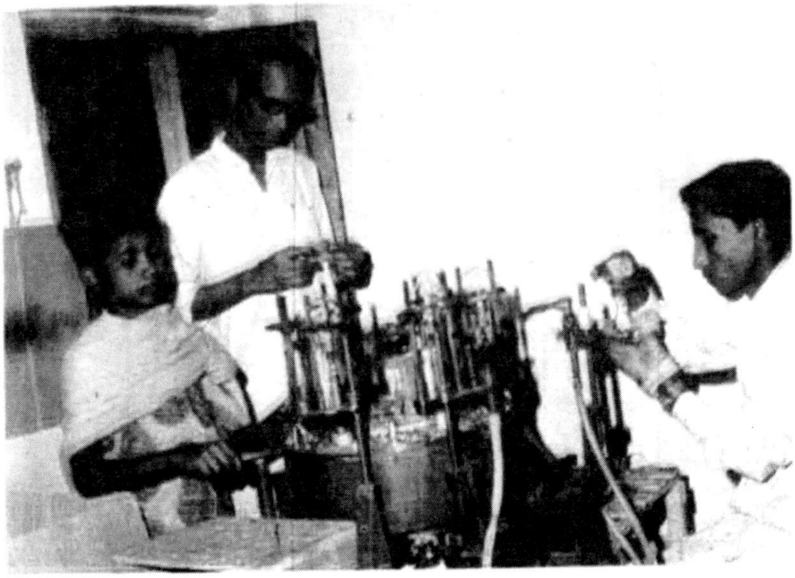

Photo 7.11. Sealing with semi-automatic machine
(Courtesy : Kerala Lamp Mfg. Co., Taliparamba, Kerala.)

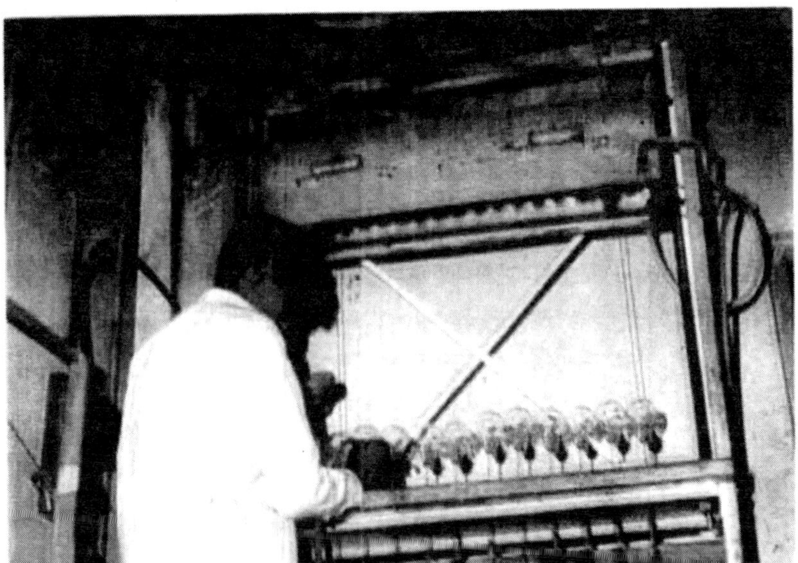

Photo 7.12. Exhausting operation in a small scale factory.
(Courtesy : Vishal Auto Lamps, Kanpur, India.)

The rotor *R* rotates eccentrically inside the cylinder with the power from the electric motor. The inlet *I* sucks air from the lamp which enters the pump cylinder '*A*' and goes out through the outlet '*O*'. The rotor has generally 3 to 4 nos. spring-loaded vanes which move up and down inside the groove of the rotor by keeping constant touch with the cylinder wall. (Fig. 7.29)

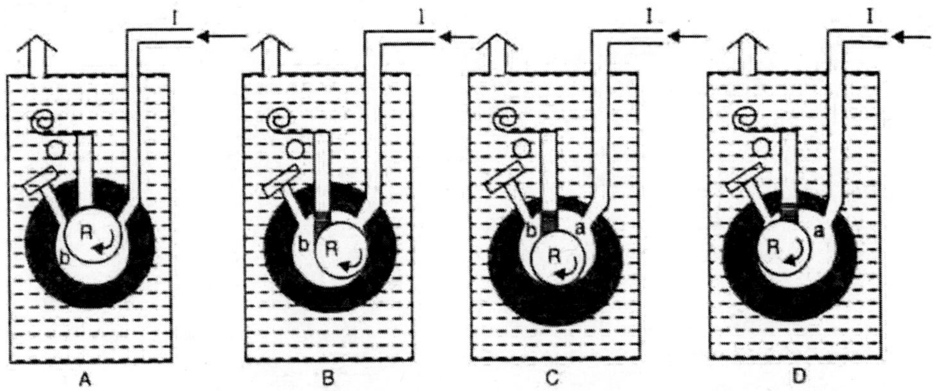

Fig. 7.29. Sequence of operations of a vacuum pump.

A. Shows the arrangement inside the pump. (Fig. 7.29)

B. One of the vane has divided the space inside the pump into two parts so that the air in (A) cannot come in contact with the air in (B).

C. As the rotor moves in clockwise direction, the air inside the space (B) gets compressed. Also at the same time air enters through (A) as suction is created by the eccentric displacement of the rotor along with the vane.

D. On further movement of the rotor, the air inside (B) gets sufficiently compressed to lift the outlet valve (*O*) expelling the air outside.

The continuous running of the motor causes more and more air to be withdrawn from the lamp creating a vacuum in the order of 10^{-3} mm of mercury.

GLS Lamps upto 100 watt are generally mass produced by automatic transfer chains. From 150 to 1500 watt lamps are produced in semi-automatic machines or on single head power aided machines. Higher wattage and some special lamps need multi-stage high vacuum pumps, which are very expensive. Mercury Diffusion pumps can be employed to overcome this difficulty.

The production of higher wattage lamp is limited and heavy investment with scanty use of an expensive high vacuum pump is undesirable for it. With little ingenuity a mercury vacuum system can be connected to the already available medium vacuum rotary oil pump.

Mercury Vacuum Pump

The principle of this pump is that in a closed circuit, the mercury vapour while passing with considerable force creates partial vacuum and when the vapour is condensed from vapour to liquid, the vacuum becomes positive and high. A continuous flow of such nature builds up real high vacuum.

The flask (F) contains mercury. A small gas heater or electric is provided to heat up the mercury slowly. The flask neck (T) is shaped as a 'V' bend, the end of which is shaped as a jet, the jet portion of the tube is enclosed inside the vacuum chamber (V). The vacuum chamber has a connection at the top to suck in air from the lamp, the lower end is connected to the rotary pump. The bottom end of the vacuum chamber is shaped as an 'U' and the complete vacuum flask (F) with this chamber are enclosed in a chamber of cool running water. This 'U' portion should always remain filled with mercury.

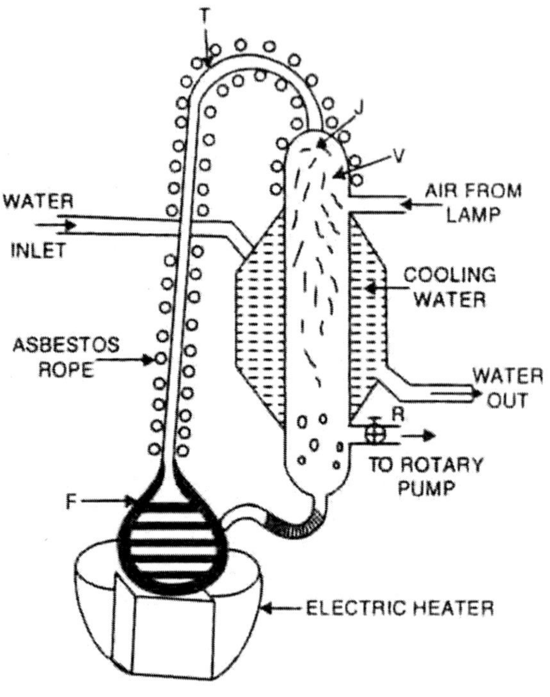

Fig. 7.30. The working principles of a mercury diffusion vacuum pump used for obtaining high vacuum. (This pump is used in conjunction with oil immersed vacuum pump.)

Then the mercury is heated slowly, it vaporises and passes through the jet 'J' as a spray. This causes depression inside the chamber and when the vapour condenses at 'C', high vacuum is created inside the chamber which draws out air from the lamp. Now the air from the lamp gets accumulated inside the chamber and by operating the valve 'R' from time to time, the same is expelled out by the rotary pump. The following table gives the characteristics of a typical mercury vacuum pump.

Edward's Diffusion Pump Series 4

Heat input—250 watt to 500 watt	
Mercury charge—50 cc	
Backing pressure mm Hg—0.1–0.25	Ultra vacuum obtained 10^{-6} torr
Speed in liters per sec in 30 ltr/sc.	
Weight of the unit—7 kg	

The mercury vapour pump acts very slowly and it is to be kept operative always. The arrangement for connecting it to the lamps and to the rotary pump should be such that at any time they could be isolated from the circuit.

After knowing about the measurement of vacuum and the vacuum pumps, further point for 'exhausting' are now discussed. While exhausting, utmost care is to be given for heating of the lamps on the exhaust bench. Unless properly heated the life of the lamp will be very low. Heating is used for vaporising the residual gas particles sticking to the inner surface of the lamp. Generally, vacuum lamps are heated for a longer period of time than the gas filled lamps. After reaching clear first stage vacuum, the lamps are to be filled with pure nitrogen gas and put to final heating, this is known as "flushing" of the lamp. In the automatic machines the adjustment for heating is not a problem but in semi-automatic benches, the timing is determined by the operator who needs proper guidance and vigilance. Water vapour is the greatest enemy of the lamp makers. By proper heating and highly efficient exhausting this is to be dislodged from the internal space of the lamp. The serious blackening which occurs due to presence of water vapour actually produces a cyclic reaction in which atomic hydrogen plays an essential part. The water vapour molecules coming in contact with the hot filament produce a volatile oxide of tungsten and the remaining hydrogen is liberated in atomic form. The volatile oxide deposits on the bulbs where it is reduced to a metallic state of tungsten by the atomic hydrogen and the oxygen of the oxide gets liberated again to act with hydrogen to form the original water vapour. This causes the same reaction to be repeated on and on. So a minute quantity of water vapour may cause a relatively enormous amount of tungsten to be carried to the bulb shell. The vacuum operator should be told about this phenomenon.

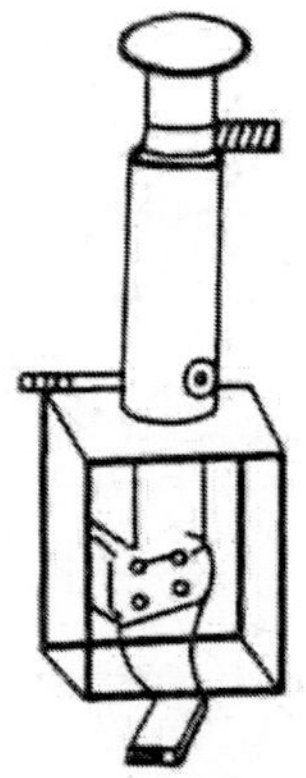

Fig. 7.31. Edwards diffusion pump a compact unit.

During exhausting, 'flashing' of the filament is also to be done. This is done when the lamp is fully evacuated and the high frequency Tester produces yellow glow on the stem pinch. For GLS lamps with 240 volts filament, a spark for a fraction of a second with 150 volt is given which ignites the getter. The flashing of filament is carried out during flashing operation of the lamp. In small factories, argon-nitrogen gas mixture is used for flashing the lamp which is an expensive procedure. This can be easily done with pure and dry nitrogen which is cheaper. "flushing" and "flashing" are two separate operations.

The next job in the exhaust bench is 'gas filling'.

Gas Filling

Evacuating a lamp and then filling it with some non-reactive gases for tungsten and lead-in-wire materials was first introduced by Langmuir in 1913. It has been found that the gas in use should have the following qualities.

1. It must be non-reactive to the internal components in a lamp.
2. It should be chemically stable at all temperature.
3. It should not create a 'flash' between the lead-in-wires.
4. It should supress evaporation of the filament.
5. It should be a poor conductor of heat.

Inert gases only fulfil all these qualifications. At first nitrogen gas was used but at high temperatures, it could not supress evaporation of the filament due it's lowest molecular weight. Nitrogen being dia-atomic in nature it has the quality to supress ionisation which causes sparks or electrical shortings between the lead-in-wires. Search was made amongst the best and cheapest gas found in the air around us which consist of the following :

Gas	*Percentage by weight*	*Molecular weight*	*Remarks*
Nitrogen	78.0	28.0	Besides these gases a small quantity of hydrogen, ozone etc. are also present in the atmosphere.
Oxygen	20.95	32.00	
Argon	0.93	39.9	
Neon	0.00182	20.5	
Helium	0.00052	4.0	
Krypton	0.00011	83.8	
Xenon	0.000009	131.3	

From the above distribution chart, it can be ascertained that zenon and krypton would be the most suitable 'lamp gas' as these have heavier molecular weight and also are bad dissipator of heat. But these two ideal 'lamp gases' are present in very minimal quantities and are obtained through expensive processes. Still these are used in special lamps as krypton gas is used in minar's lamps and xenon gas is used in cinema projector lamps and floodlight lamps for lighthouses. Commercial production of these two gases have not yet started in India.

For GLS lamps and auto lamps, the best and the cheapest gas was found to be argon. Because of it's heavier molecular weight than nitrogen, it conducts heat slowly towards the outer shell at the same time turns out to be a bad carrier of evaporated tungsten. At higher temperatures, nitrogen reacted with tungsten and tungsten molecules got seggregated from the filament which got deposited on to the shell wall. With the introduction of argon, these difficulties were overcome. More concentration of heat was now possible as there was no chance of reaction and the 'coiled coil' filament was introduced showing very high performance rating of the lamp. This was a great achievement but all was not well with argon filling. At high temperatures, argon gets ionised and becomes a good carrier of electricity. It starts internal short circuit and suddenly destroys the lamp without any easily detectable cause. In India every year it had been found repeatedly that the supply of argon-nitrogen gas mixture is totally stopped for about two to three months. During this period all factories use either pure argon or nitrogen as lamp gas. Lamps of higher wattage have been found to be filled with only argon. The life of the lamps made in this manner will always be in question and this practice is not to be encouraged at all.

The ideal lamp gas is a mixture of argon and nitrogen. At higher temperature the nitrogen molecules form a barrier for the ionised argon gas at the low density zone near the lead-in-wires and avoids the destructive spark. This spark could be also avoided by creating positive pressure inside the lamp but there will be then possibilities of violent explosion endangering the lives of the users. Technically perfect in all respect, lamps can be made repeatedly in batches after batches only with fully automatic machines under strict control. More heat from the filament can be used in these lamps. For this reason, the makers use high-argon less nitrogen gas for these lamps. For technically semi-perfect lamps it is advisable to use mixture with about 15% to 20% nitrogen.

In the linear type of halogen, krypton or xenon lamps pure gases are filled though these gases also have these deficiencies. There is no chance of sparking in these lamps as the electrodes are placed far apart at the opposing ends of the tubing and also the gas is filled in at high positive pressure.

Nitrogen is cheaper than argon gas. In India more lamps are made with 86% argon and 14% nitrogen gas mixture. With this combination a little money in nitrogen is saved but in the long run this proves to be uneconomical. Nitrogen at higher temperature acts as a carrier of tungsten molecules which shortens the life of the lamp. If more nitrogen is filled the pressure inside the lamp also should be increased which is again not desirable. The pressure inside the lamp in cold condition is around 710 mm of mercury and it is not desirable to exceed this limit in GLS or Auto Lamps.

The proportion of mixture of argon and nitrogen depends on voltage rating, filament construction, temperature and the spacing between the electrodes at the stem-pinch. The following gives an universally used optimum proportion of mixture :

6 volt lamp—Pure argon with trace of nitrogen

12 volt lamp—Argon 86% and nitrogen 14%

240 volt lamp—Argon 50% and nitrogen 50%

without fuse type lead-in-wire

110 volt lamp—Argon 93% and nitrogen 7%

with fuse wire

The 92/8 argon/nitrogen gas mixture available in India is good combination for GLS and high wattage auto lamps. It will be useful here to note the following chart showing the fruitful use of energy in a lamp and the amount which goes wasted. The figures are expressed in percentage of the wattage of the lamp.

Wattage	*Type vacuum or Gas filled*	*Filament used single/Coiled coil (C.C.)*	*Radiation as visible light*	*Total filament radiation beyond lamp*	*Total gas loss*	*Base and bulb loss*	*End loss*
1	2	3	4	5	6	7	8
6	Vac	Single	6.0	93.0	—	5.5	1.5
25	Vac	Single	8.7	94.0	—	4.5	1.5
40	Gas	C.C.	7.4	71.3	20.0	7.1	1.6
60	Gas	C.C.	7.5	80.8	13.5	4.5	1.2
100	Gas	C.C.	10.0	82.5	11.5	5.2	1.3
200	Gas	Single	10.2	77.4	13.7	7.2	1.7
300	Gas	Single	11.1	79.8	11.6	6.8	1.8
500	Gas	Single	12.0	82.3	8.8	7.1	1.8
1000	Gas	Single	12.1	87.4	6.0	4.7	1.9

In the foregoing chart, column 4 gives the most important figures showing the percentage of energy actually received for use as visible light. For a 100 watt lamp, only 10 watt is available as visible light, 11.5 watt (col. 6) of energy is lost by the filament due to conduction of heat by the gas around it. Even the loss due to the moly pig tails and the lead-in-wires is 1.3 watt.

Column 5 gives the total radiation. For one 100 watt lamp, it is seen that 82 watts are obtained as total radiation and the rest of the 18 watts are lost through glass bulb and lamp cap and out of 100 watt only 10 watt serves the user.

The efficiency of production reflects on columns 6, 7, and 8. A simple defect as over gettering will increase the loss in column 7 and the result in column 4, will become less.

These figures give a clear idea to the manufacturer as to how less a consumer receives against so high an input. So, the maker should be ever vigilant to bring perfection in his manufacturing process and improvement in production to make lighting expenses more and more economical.

Drying of Argon-Nitrogen Gas Mixture

The commonest and the most destructive contaminant in a lamp is the presence of oxygen molecules either freely or in the form of water vapour. In the exhaust bench the lamp shells are heated for a considerable period for evacuating the lamp of any gas or vapour. There again remains the risk of water vapour when argon-nitrogen (lamp gas) is introduced inside the lamp. The manufacturers of lamp gas warns of the presence of traces of water vapour in the cylinders so it is always advisable to dry the argon-nitrogen mixture by passing it through silica gel or phosphorus pentoxide to make the lamp gas free from water vapour. Generally P_2O_5 (phosphorus pentoxide) is used for this process keeping in view of the following points :

1. The circuit should be made completely leak proof for which araldite, bostik and other joining compounds are to be used liberally.

2. The glass towers should be made of pyrex or corning glass.

3. Two foster valves are to be installed; one at the starting of the lamp gas inlet near the gas cylinder and another after the tower near the bench to avoid over pressurising and bursting of the rubber connections in the circuit.

4. The tower near the pump suction side to be kept free of any chemicals and is to be filled with small pieces of clean and dry glass rods tubes and glasswools.

5. The P_2O_5 powder must have sufficient open space on its top to fly up and spread down.

6. Under no circumstances the P_2O_5 should be permitted to enter the vacuum manifold.

Phosphorus pentoxide is a very quick absorber of moisture from the atmosphere. While filling this into the glass towers, great care should be taken to avoid its contact with any moist surface. Its exposure to the atmosphere also is to be minimised. The circuit should be made inside a hot dry room. The Argon-Nitrogen Circuit (ANC) generally consists of 3 or 4 corning/pyrex glass towers of 3 feet length and 3 or 4 inches inner diameter. The top end of the tower has a neck for inserting materials which is better to be sealed by melting. The tower has one opening at one side at the top and one near the bottom where rubber tubes are connected to provide passage for the gas to flow.

Photo 7.13. 24/36 head automatic seal/exhaust machine.
(Courtesy : Badalex—A Div. of Sale Tinley Tech., England.)

Photo 7.14. Automatic filament feed unit.
(Courtesy : Badalex—A Division of Sale Tinley Technology, England.)

Photo 7.15. Mounting conveyor chain.
(Courtesy : Badalex—A Division of Sale Tinley Technology, England.)

Photo 7.16. 60 station automatic capping machine.
(Courtesy : Hindustan Machine Tools Ltd., India.)

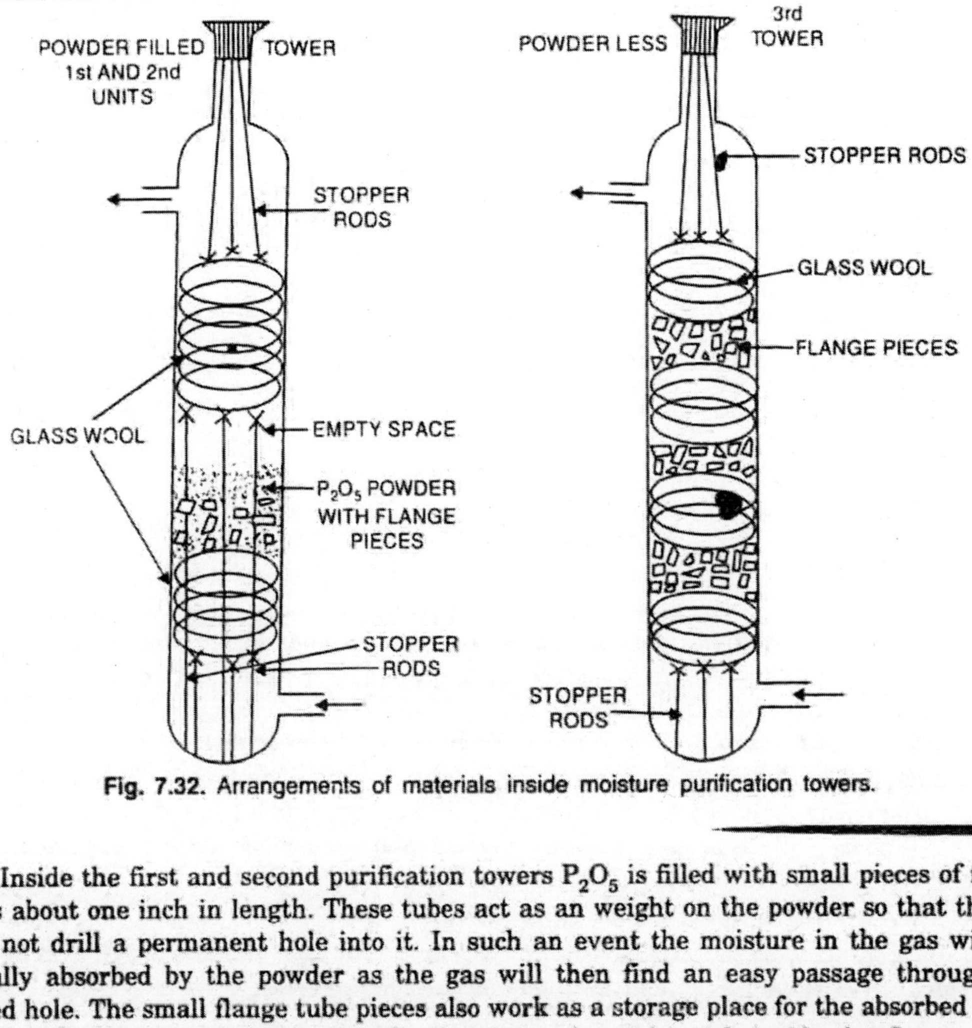

Fig. 7.32. Arrangements of materials inside moisture purification towers.

Inside the first and second purification towers P_2O_5 is filled with small pieces of flange tubes about one inch in length. These tubes act as an weight on the powder so that the gas does not drill a permanent hole into it. In such an event the moisture in the gas will not be fully absorbed by the powder as the gas will then find an easy passage through the drilled hole. The small flange tube pieces also work as a storage place for the absorbed water vapour which gets coagulated with the powder and mostly stick inside the flange tubes. Unbroken solid rods are used as separators for the glass wool on the ends of which a few pieces of small glass rods are welded to avoid slipping down of the glass wool.

The glass wool should not be packed so tightly that the gas will not be able to pass through it or it should not be so loose that the P_2O_5 powder will pass through it. The glass wool actually acts as the filtering agent. The compactness of the glass wool is achieved by making about one inch thick strand of thread. The empty place above the powder is kept to provide splashing down place for the powder forced upward by the gas entered through the opening at the bottom. If there be any moisture, the powder shall immediately hold it in suspension.

The third tower is kept free of powder and is filled with four or five rings of glass wool and small pieces of solid rods and exhaust tubes. Practically, it is seen that the first tower catches the maximum moisture and the moisture which by passes the first tower is caught by the second tower. In case the second tower also is seen to absorb more quantities of moisture then one more tower with P_2O_5 will be essential. The more are the number of towers the more time it will take for cleaning or evacuating the circuit and there will be more chances of vacuum leak from joints. Generally two towers with P_2O_5 and one free tower are used.

ANC CIRCUIT MAKING

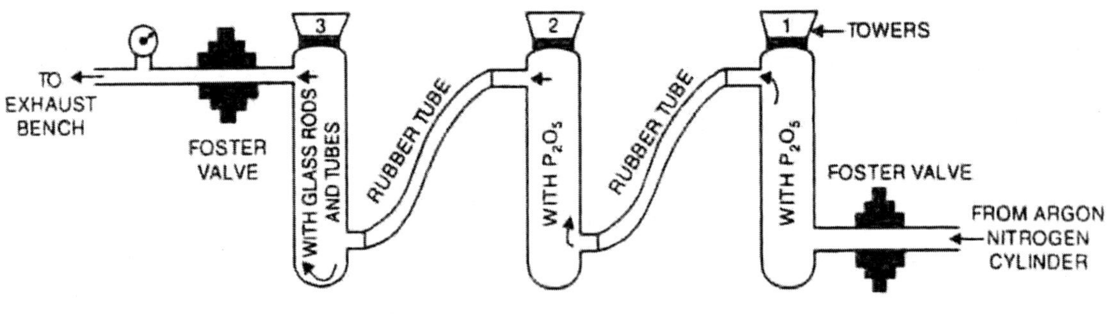

Fig. 7.33. Connections of purification towers.

Great care should be taken when towers are to be 'filled in' newly. It should be done on a hot dry day in a closed room. For making the room moist free, two smokeless ovens or electric heaters are kept burning in this room for one hour before starting actual filling. All the towers, glass pieces, are to be washed and dried over flames. After completely drying the towers over gas flames, plug-in the openings with rubber corks, and let it cool down in room temperature. Make the work place ready with all the necessary components placed near at hand. The work on the third or the empty tower is done first as it does not contain any P_2O_5 and any moisture going inside it would be very difficult to remove. After working with P_2O_5 there will be always moisture present and some of it surely can find its way if the third or empty tower is done at the end.

While working with glass wool and P_2O_5; good care is to be taken so that these do not come into contact with the skin. Everyone present should wear full sleeved clothes to absorb the moisture of their bodies and also should wear skin gloves.

The act of filling P_2O_5 should be done very fast. Both the glass wool packings must cover the bottom and the top openings of the tower. To make it a completely leak-proof unit the top of the neck is welded which is not necessary if the work is done in a proper manner. The neck of the tower can be also very tightly closed with a solid rubber stopper so that the internal pressure of the gas should not be able to open it. Then molten wax is filled in the cup, even covering the rubber stopper. As told earlier all the joints are to be treated with best quality resin compounds for any probable leakage. For all connections use high pressure rubber tubes.

At the beginning, there should be 8 nos. 100 watt lamps around the tower to heat them up. For everyday use, four lamps are to be kept burning always.

For the first evacuation of the circuit, seal all the heads and obtain perfect vacuum for the pump and the manifold. Then evacuation of the circuit is to be done in stages. Isolate the third tower from rest of the circuit by pinching the rubber connection between the third and second tower. When proper vacuum colour is obtained by the electrostatic tester, the second and the first towers also are to be evacuated in the same manner. For the first time, it might be necessary to heat up the second and the first towers containing P_2O_5 with carburising flame from the burner. This is necessary to release the water moistures which got lodged on to the inner wall of the towers during the filling process.

The towers are better to be installed inside a box with detachable side panels. Argon-nitrogen circuit should never be left in vacuum condition. This should be always filled with gas. When there is indication of yellowish slurry deposit at the bottom of the powder, the circuit should be dismantled and remade.

Exhaust Tube Tipping. This is the final job on the exhaust bench. Some manufacturers prefer to keep the parting off point as a fine tip and the others prefer to bend it at the molten stage. It is alleged that the bent tips are prone to slow leaks at the same time the fine tips give away very easily on shocks or while capping if it is cut a little longer. There might be truth in both the claims. It appears that the small lamps with fine tips give better service and the large wattage lamps producing more heats will prove better with bent tips.

Operation 13

Sealing and Exhausting Fully Automatic (*See Photo 7.13*). The sealing turret and the exhausting turret are combined on one machine. The sealing turret is equipped with 24 heads which are rotated by the drive mechanism provided by a separate electric motor controlled independently of the main drive. The sealed lamps are automatically transferred to the 36 heads exhaust turret via small multi-station index turret and during this transfer the lead wires are spread out so that they are clear of the burners when the exhaust tube is melted. The mounted stems are fed from the mount conveyor automatically into the head spindles of the sealing machine. Near to the same station, the shells are brought by the bulb washing conveyor which feeds these directly into the head cradles over the mounted stem.

After the shell has become sealed on to the flange of the mounted stem, the skirt is cut away and the neck of the bulb is further heated to shape by closing of the jaws with the support of air injection. The neck and the seal are annealed after shaping and then transferred to the exhaust turret.

The 36-head exhaust turret is equipped with compression rubber type heads with closure clips which automatically close and open the heads and provision is made for automatic leak detection. The connection to the head carrying a leaking lamp from the valve plate is automatically closed ensuring that the vacuum manifold is not adversely affected. Any bad lamp is automatically removed from the turret prior to the transfer action.

The early stations of the exhaust turret are covered by an oven while the rest of the stations are cooled by air jets. The pumping schedule can be adjusted as per requirement.

The lamps are filled with gas through regulated control valves. The exhaust tube is tipped off and the stump removed from the compression head automatically. The tipping off is achieved by the pressure tip-off process.

The machine is capable of making 3000 lamps per hour.

Station Layout of 24-Head Sealing Turret :

1. Mount lead
2. Stem (detect-no stem, no bulb)
3. 5, 21, 22, 24–blank
4. Bulb load (detect no bulb stem blow out)
6. Seal burner brand burner
7, 8, 9. Seal burner brand burner
10–13. Seal burner
14–15. Cut off burner

16–17. Head up burner

18. Neck mould

19–20. Anneal burner

23. Transfer to exhaust turret via a small intermediate multi-station indexing turret.

Station Layout of 36-Head Exhaust Turret :

1. Load lamp, close compression head
2. Wire setter and rough vacuum
3. Rough vacuum
4–5. Blank
6. Leak detect and re-vacuum
7. Pump
8. Nitrogen flush
9. Blank
10. Pump
11. Nitrogen flush
12. Blank
13. Pump
14. Nitrogen flush
15. Blank
16. Pump
17. Nitrogen flush
18. Blank
19. Pump
20. Nitrogen flush
21. Blank
22–29. Pump
30. Pump, preheat burner
31. Argon fill, preheat burner
32. Balance, 1st tip off burner
33. 2nd tip off burner, detect leaky lamp and take out
34. Unload into intermediate turret which also inverts the lamps before loading into cap threading conveyor, stamp extraction
35. Blank
36. L.P. head cleaning.

Note : (*a*) Stations 3 to 10 are covered by oven.

(*b*) Stations 11 to 30 are covered by air jets.

Utilities required.	Fuel gas	0.14 kg/cm^2 (2 lb/sq. in.)
	L.P. air	0.28 kg/cm^2 (4 lb/sq. in.)
	Oxygen	0.35 kg/cm^2 (5 lb/sq. in.)
	Vacuum pumps...medium vacuum (Rough)	

Filament Feed Unit—Automatic Machine. This unit has been designed for the automatic feeding of coiled coil filaments on to the Auto Stem and Mount (AUTOMOUNT) Machine (*See Photo 7.14*).

A batch of filaments is loaded into the vibratory bowl feeder and the filaments are automatically separated and loaded singly into specially designed carriers. The carriers are supported on a continuous belt and are conveyed to the filament loading station of the mount mill. If a carrier remains empty, then an inhibitor prevents the transfer of a stem into the mount mill so that when the empty carrier arrives at the filament pick up station the head at that station also be empty.

All utilities for this unit is derived from the automount.

Mounted Stem Conveyor—Fully Automatic (*See Photo 7.15*). This conveyor is a linking unit between the two automatic manufacturing machines *i.e.*, the automount and the seal/exhaust machines, which run at different speeds to complete their operations. It is essential that a head on the conveyor is in correct position to receive a mounted stem from the automount and at the same time another head on the conveyor is to be so placed that a mounted stem can be transferred directly into the seal/exhaust machine. These operations are assured by two separated indexing mechanisms placed at the ends. The conveyor receives its service utilities both from the automount and general service system.

Operation 14

Ageing (G). The cap-less lamp is received from the exhaust bench now to be qualified of its light and life. This is the first working point where the decision about a lamp is taken.

Ageing is necessary for long and high voltage filaments as used in GLS lamps. For mini and auto lamps, the voltage being low it is not necessary to age them.

Ageing means the gradual raising of the filament temperature to avoid shearing off due to sudden increase of voltage. In other words it is the gradual normalising process of tungsten molecules so that these can occupy a permanent place with the new burnt out condition of the tungsten when current is passed through it for the first time. Ageing should be carried out very smoothly and gradually increasing the voltage in stepless, sparkless stages.

All GLS lamps need be aged not for its length of the filament only, it needs ageing as the filament is always fully gettered and this has got to be burnt out in a proper manner so that the getter smoke is recessed mostly inside the cap. Ageing is carried out with base up position.

The voltage variations or stepping up action should be made gradual. It should be operated by a stepless sliding contact switch so that there is no sparking or jumping while changing voltage.

By flashing in the exhaust bench the unwanted getter and other foreign particles have already been removed from the lamp. In case flashing was not done, the ageing must start from the flashing point of the getter. For 240 volt vacuum filament the initial point may be kept as 60 volt and the rest of the range being increased in steps of 25 volts; as for 240 volt gas filled filament, the initial starting voltage should be 150 volts and the rest of the range increased in steps of 25 volts. Vacuum lamps are to be aged slowly and the gas-filled lamps are aged comparatively faster.

At rated voltage, the lamp must be allowed to burn for a longer time and at the end it is to be fed with 10% extra voltage for a short period. The lamps with faulty mounting or vacuum leaks gives away at this point.

Operation 15

Capping (G.A.M.) (*See Photo 7.16*). The process of fixing the metal cap on to the neck of the 'aged lamp or cap-less lamps' is known as capping. For mass production and easiness of the job capping is mostly done in automatic machines. Where utmost care is to be taken to maintain a predetermined gap between the cap and the lamp, this is better done in the heating chamber. For fixing caps to small lamps even hot sand is used. The main criteria in capping is the proper curing of the capping cement.

The lamp caps are classified as following :

(*i*) Bayonet cap (*B*) used on *GLS* lamps

(*ii*) Bayonet cap automobile (*BA*)

(*iii*) Bayonet caps automobiles with off-set pins (*BAY*)

(*iv*) Edison screw cap (*E*)

(*v*) Prefocus (*P*)

(*vi*) Single contact (*s*)

(*vii*) Double contact (*d*)

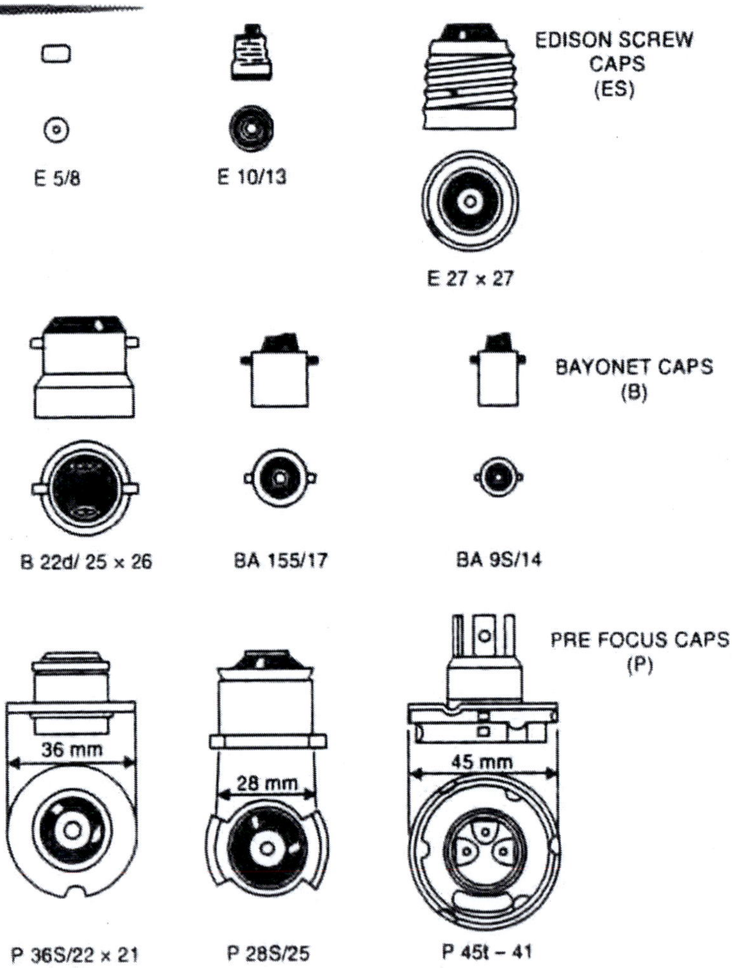

Fig. 7.34. Metal caps in use with mini, auto and GLS lamps.

The lamp caps have a collar to accommodate a thin layer of cement which acts as the bonding material. If the cement is carelessly applied or allowed insufficient time and temperature for setting, bad capping will be the result. It is desirable to have a continuous glue line of cement as thin as possible to achieve the best result. It should be understood that the strength of a properly bonded joint made with thin layer of cement whether under tension or shear will be always stronger than the glue put in bulk.

Since the cement becomes weaker with age and particularly with heat and humid working conditions, some heat resistant special system or wedge based lamps are now being introduced.

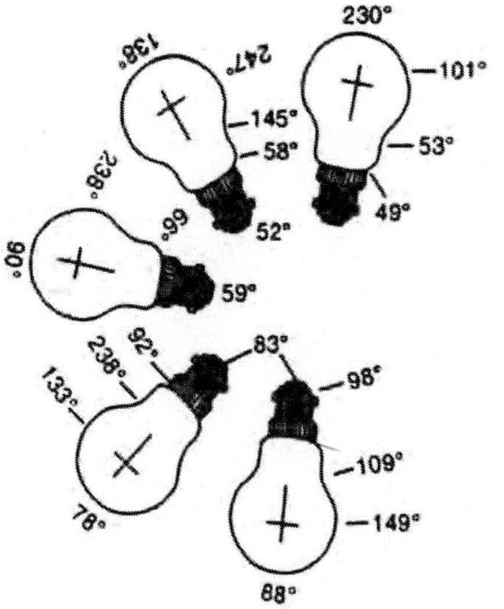

Fig. 7.35. Temperature variations in a lamp at different burning angles for a 100 watt lamps.

Generally the cement available is meant for lamps with low lamp cap temperature. In the small factories, these cheaper cement is used for head lamps or GLS lamps above 60 watt lamps which get their caps loosened after a few working hours. For all high wattage hilamp cap temperature lamps, the capping cement must be mixed with extra 2% to 5% silicone resin by weight.

The capping section should have a lamp gap torsion testing apparatus. A few lamps from every batch are to be tested for maintaining the following strength of their caps.

Types of cap		*Torsional moment*
E–27, *B* 22	....	30 kg/cm
E 40	....	50 kg/cm
BA 7, *BA* 9, *E* 10	....	2.5 kg/cm
BA 15, *BAY* 15, *BA* 20, *P* 22	....	11.5 kg/cm

For easy operations and safety, the connection of the main filament has been standardised in auto head lamps.

In Pre-focus Lamps. The main filament shall be connected to the contact which is nearer to the slot of the ring.

In BA 20d Cap Lamps. It is to be connected to the *left hand* side of the contact when the cap is viewed from below with the bigger pin at the bottom.

In BAY 15d Caps. The low wattage filament should be connected to the *right hand* contact when viewed with cap towards observer, index pin down and low wattage filament uppermost.

Cementing and capping operations are generally carried out by unskilled workers. These two operations also need good supervision. For different types of cement the heat of the capping machine burners need adjustment. Sometimes the caps are not fixed even after two rounds of heat on to the same caps. If the flame of the burners are not properly adjusted this heats up the neck of the lamp which becomes very hard and brittle. In fact the name of the manufacturer is turnished for such a simple mechanical defect as loose cap.

Operation 16

Soldering (G.A.M.). The extra length of the lead-in-wire are cut off to just protrude over the eyelet hole. The soldering man has generally 3 to 4 lamps in his hand and touches the tip of the solder and the hot soldering iron just on the eyelet body and never permits the molten solder to pass through the hole of the eyelet. Due to careless soldering molten solder enters into the flange and at times this provides an easy passage to the electric current which blows out the fuse in the house. The light centre length is measured along with the thickness of the solder. The thickness of the solder should be as per the stipulated standards. Sometimes it becomes difficult to solder the electrode on to the side of the metallised caps. A little strong killed spirit eradicates this difficulty. It is not at all advisable to continue heating of the cap when solder does not stick on to the cap body, this affects the capping strength.

For assymetric prefocus lamps special gadgets are made for final adjustment of the focus by shifting cap position of semi-baked caps.

Operation 16(*a*)

Cap Threading and Finishing Conveyor (With Ageing)—Fully Automatic (*See photo 7.17*). This is a 36 station linear indexing machine which combines the operations of cap threading, cement baking, wire trimming, soldering and ageing. Loading, unloading and all the above operations are automatic. The conveyor is synchronised with the drive of the sealing/exhausting machine. The machine automatically straightens the lead wires which were bent sideways for exhausting operation, threads them through the holes of the cement pasted caps, bakes the caps, trims the excess wires, solder the wires, ages the finished lamps and finally unloads them. The caps are cemented by a base filling machine which automatically feeds this cap threading conveyor at its station number 10. The machine indexes upto 3000 units per hour.

Utilities needed. As in sealing/exhausting machine, including vacuum arrangements needed for cap threading.

Station Layout for Edison Screw Caps (Horizontal Filaments, Unfused Electrodes)

1. Load	2. Centralise
3. Straighten lead wires	4–5. Blank
6. Spread	7. Blank
8. Push side wires down	9. Detect lamp
10. Feed and thread cap	11, 16, 45, 50–55, 74–76 All blank
12. Top wire stretch	13. Cap push down
14. Continuity check	15. Reject
17. Lower register plate	18–37. Blank
38–42. Cool cap	43. Top wire cut
44. Top flux	46. Top solder
47. Side wire puller	48. Side wire outer

Photo 7.17. 76 head cap threading and finishing machine.
(Courtesy : Badalex—A Div. of Sale Tinley Technology, England.)

Photo 7.18. Cap threading machine.
(Courtesy : Badalex—A Div. of Sale Tinley Technology, England.)

Photo 7.19. Full view of A GLS chain.
(Courtesy : Hindustan Machine Tools Ltd., India)

49. Side wire scraper
52–69. Age vertically
70. Test
71. Lift register plate
72. Reject
73. Unload.

Cap Pasting Machine—Fully Automatic. This is a feeder machine to the cap threading and finishing machine. It is used to paste the inner base of the cap with a uniform layer of paste and suply the same synchronisingly to the cap threading conveyor.

The machine is an 8 station indexing machine suitable for limited type of caps to be used. Caps are fed from a vibratory feed hopper to the pasting positions with the machine working on dual action principle. The two cement pots at the top are fitted with positive feed plungers which feed cement to the cap inner base. This has a capacity to supply 3000 pasted caps to the cap threading machine. Utility H.P. Air at 5.5 kg/cm^2 (80 lb/sq. in.)

Station Layout

1. Load two caps
2. Blank
3. Fill left hand cap
4. Blank
5. Fill right hand cap
6. Unload both caps

7–8. Blank.

(See Photo 7.18)

Operation 17

Final Testing (G.A.M.). The lamp thus produced by following all the laid down standards and observing all the precautions is to be finally tested which is more of a routine check. The capping, soldering should be good, there should be no physical defect and the lamp is ignited on a holder to check its serviceability.

Operation 18

Cleaning and Packing G.A.M. The finally tested lamp is to be cleaned at first with moist cloth. The eyelets and solders are to be very thoroughly cleaned with moist cloth. The same lamp is immediately cleaned with a clean dry cloth by another worker. This must be carried out to avoid rusting of the aluminium caps due to traces of residual moisture and chemicals used while soldering. The shell of the lamp must be made to shine and put immediately inside the packing carton filled with packing material touching the body of the lamp with the carton.

Thus a mini, auto or GLS lamp is made ready for instant use.

(See Photo 7.19)

8

Halogen Lamp Technology

What is Halogen ?

The word Halogen means a material which has salt forming qualities. The sea water is salty and the main halogen elements chlorine, bromine, iodine are available in the sea water. There are five elements *i.e.*, fluorine, chlorine, bromine, iodine and astatine belonging to group VIIB of the periodic table known as halogens. For production of halogen lamp, one or more of these elements are introduced inside the lamp which improve on the incandescence process.

Halogen lamps are still on very earnest and vigorous research stage of development. So, it is essential to know about the physical properties of all the members of the halogen group as any one of them could become a prominent donor for making best quality incandescent lamp.

Physical and Atomic Properties of Halogens

	Fluorine	*Chlorine*	*Bromine*	*Iodine*	*Astatine*
	F	*Cl*	*Br*	*I*	*At*
Atomic number	9	17	35	53	85
Atomic weight	19.00	35.457	79.81	127	—
Physical state	Gas	Gas	Liquid	Solid	—
Colour	Pale yellow	Greenish yellow	Dark red	Violet	—
Melting point °C	–223	–101.6	–7.2	113.9	—
Boiling point °C	–188	–34.6	58.7	184.4	—

The atoms of halogen contain seven electrons in the outermost orbit which very easily complete their octet by receiving one electron from their neighbouring element. This forms a monovalent negative ion. As the atomic number increases, the radius of the atom also increases and the attractive force towards the monovalent nature of the negative ion also diminishes. For making halogen lamp this characteristic of the halogen element is to be remembered. Astatine having highest atomic number will loose its negative ion very easily and become electropositive. As astatine is a rare material, iodine is the most popular halogen element. Iodine becomes electropositive very easily and stronger in positive characteristics. So, while introducing halogen compounds inside a lamp, their properties are to be taken care of.

Fluorine

Fluorine occurs in nature abundantly as fluorite, and cryolite (sodium, fluoroaluminate). calcium fluoride is one of the essential minerals for our teeth. Fluorine is mostly used in very high voltage insulation as industrial reagent, manufacturing of freon gas for refrigerator etc. Fluorine is the most reactive of all halogens. In lamp industry hydrofluoric acid is used specially for frosting the inner side of the lamp shell. It shows how difficult it will be to introduce fluorine inside a halogen lamp. Fluorine attacks glass as well as tungsten in a lamp. If these two disadvantages could be overcome it would have the most suitable properties to prolong the life of tungsten filament lamp.

Chlorine

In sea water, rock salt, chlorine is found as sodium chloride. Chlorine is a pale green colour gas having suffocating odour. From our daily use, we see bleaching powder being used in paper and cloth industry. In lamp industry, we use hydrochloric acid for washing shells. Chlorine by products are most essential materials for our bodily functions. A large quality of medicine base is hydrochloric acid. This attacks glass but also has the quality to improve the quality of lamps.

Bromine

The main source of supply of bromine is sea water. This halogen is found in sea water only 60 to 70 parts per million. The recovery of bromine is very cumbersome and expensive process.

Bromine is a dark red liquid with suffocating odour. Bromine vapour has very irritating action on nose, eyes and throat. Care must be taken to handle the liquid as it causes severe burns on skin.

The molecular weight determination of bromine shows that in the vapour state, the molecules are diatomic but when the gas is heated, dissociation takes place beyond 1050°C.

Bromine is soluble in water but is a powerful oxidising agent. Bromine combines easily with sulphur, phosphorus, arsenic and antimony.

Bromine has already established its place in halogen lamp making.

Iodine

In sea water iodine is found only one part per million. Also iodine never occurs free in nature. Presently iodine is extracted from deep sea weeds and caliche (saltpetre from chile). The other source being the oil brines found in deep oil wells. These oil brines sometimes contain 60 to 70 per million iodine.

Iodine is a grey black solid having a metallic lustre. It crystallises in plates. The odour of iodine has its distinct characteristics.

When heated quickly, the iodine sublimes without melting but on gentle heating to 114°C, it melts and starts boiling at 184°C.

Upto the temperature of 700°C, iodine retains its diatomic physical nature I_2. At the further rise of temperature. I_2 breaks up as 2I becoming monoatomic molecules. These properties of halogen are being utilised in halogen lamps.

Iodine freely dissolves in carbon disulphide, chloroform and benzene. It is also very soluble in aqueous solution of potassium iodide.

For purifying iodine, grind it with about half its weight of potassium iodide which will remove the chlorine or bromine present with it as impurities and sublime off the iodine. Remove the sublimate and resublime as above without addition of potassium iodide. The re-sublimed iodine is then put into a dessicator containing anyhydrous calcium chloride or phosphorus pentoxide (P_2O_5) and let it stand overnight for removing any water vapour.

The present halogen lamps are made by introduction of iodine and it is essential to have thorough knowledge about all the halogens particularly iodine and bromine which are used extensively.

Astatine

The heaviest of the halogens and is a very rare element. Astatine finds its use only in laboratories. There is no commercial usage of astatine and has not been tried in the lamp industry because of its non-availability and unexplored chemical properties.

The Regenerative Cycle

A tungsten filament lamp is a constant potential radiation source. The lumen value of such lamps reduce gradually due to evaporation of tungsten filament. Actually the heating process inside the lamp due to the electrical pressure and heat cause the tungsten molecules to get seggrigated from the coil and collide against the molecules or argon nitrogen gas surrounding the filament. Some of the molecules though try to settle back to the filaments, a majority of them fly off and get themselves deposited on to the shell wall. In 1913, Langmuir introduced inert gas inside a lamps to create a pressure around the filament to prevent evaporation loss. To have maximum efficiency of light, heat must be maintained continuously inside the lamp. To achieve that temperature, the molecules on the surface of the tungsten filament also should not be over worked. The heat loss inside the lamp is due to conduction which means that the molecules inside the lamp receive and pass on the heat quantas to the adjacent molecules till they reach the glass shell and there is no flow of hot gases inside the lamp. From the glass shell the heat is dissipated by convection process to the atmosphere around it. So, this inert gas inside an incandescent lamp works as a covering or a sheath. The dimension of this sheath is of utmost importance while designing a lamp. The theory of 'regenerative cycle' is based on this sheath. This is "Langmuir sheath" named after its inventor. While designing a GLS Lamp, we have the filament placed at the centre of the globe so that the heat produced by a particular wattage can be used to its maximum useful value. Since the beginning of this century several attempts were made to arrest the evaporation of tungsten filament by working on the 'Langmuir sheath' *i.e.*, the atmosphere around the filament by using different inert gases, chemicals, shapes and filament materials, so that optimum light value could be obtained with minimum losses of tungsten.

Earlier Theoretical Concept of Regeneration Cycle

In halogen lamp the word 'regeneration' relates to the process of 'bringing back' the tungsten molecules which got separated from the main body of the burning filament. In the twenties, it came to be known that at a certain temperature range, chemical reactions take place between tungsten and halogen gas molecules and there are indications of redepositing of the 'thrown off' tungsten back on to the filament body.

Due to non-availability of proper materials and implements, this remained under experimental observations for nearly 30 years. Then around 1959, it came to be known that by using a tube of non-crystalline quartz glass, a linear type of lamp could withstand an wall temperature of at least 250°C which would be suitable to support the regenerative cycle of tungsten-halogen-tungsten. A measured quantity of "iodine" (one of the halogen) with pure Argon (inert gas) as carrier were introduced inside the linear lamp and a successful regenerative cycle was achieved. By 1962 there were many firms all over the world offering halogen lamps for flood lighting, automobile lighting etc.

All halogens—iodine, bromine, chlorine and fluorine can react with tungsten and start the tungsten-halogen regenerative cycle. In the illustration, iodine is taken to show the process of regeneration combination with pure argon gas. Depending on the cross-section of the tungsten filament, this operates at a temperatures between 2600°C to 3000°C. As described earlier, the Langmuir sheath, or retention of heat inside the lamp, plays the main role in this regeneration cycle. The heat inside the halogen lamp divides itself into three distinct zones where different processes take place.

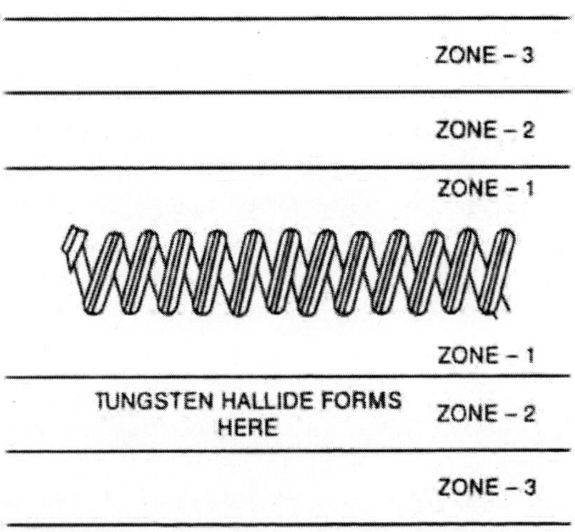

Fig. 8.1. Halogen regeneration process inside a quartz tube.

Zone 1. In the lamp under illustration, iodine stays in vapour state in mixture with pure argon. This zone is the hottest zone forming a thin layer surrounding the tungsten coil. At this point the working temperature is 2600°C or above. The bonding of the tungsten crystals are loosened here and the tungsten molecules start flying off from the filament. Also due to the excessive heat in this area the tungsten vapour, iodine vapour and argon gas all remain as separate entities without any interaction between them.

Zone 2. After the first zone comes this area which plays the most important part in regeneration cycle. Here the Iodine becomes a carrier. In this zone, at temperatures between 1000°C to 1600°C, the iodine combines with the tungsten and forms tungsten halide.

Actually this zone has two sections. The area of this zone adjacent to zone 1, being hotter, the halides get disintegrated as iodine and tungsten.

The area further away from zone 1 produces the halides with tungsten. This is the most important zone in a halogen lamp.

Zone 3. This zone covers the inner surface area of the glass tube. From zone 2, this being sufficiently cooler, here we have tungsten halide argon gas and the iodine gas which are just in excess of available tungsten molecules. The shell temperature being at 350°C to 500°C, the tungsten halide or iodine cannot get embeded on to it because of the heat.

As told earlier, the most important zone is the second zone. Here at one side, tungsten halides are formed at colder region and tungsten halides are broken by splitting into tungsten and iodine in the hotter zone.

In actual process, soon the desired temperature is reached and the first batch of tungsten molecules are thrown out. Tungsten halides also are formed and at the same time they are splitted into tungsten and halogen. This chemical reaction and reverting process not only produce a halo or pressure on the throwing off behaviour of the tungsten from the filament but the thrown out tungsten liberated from tungsten halide actually returns to the filament and stops its working decay. At the same time, the iodine of the tungsten halide is again made available to restart the cycle.

From the temperature gradient seen inside a halogen lamp, it is now clear that the glass covering or the shell has got to be at a measured distance from the centre axis of the filament. Any increase in diameter of the linear shell will place the three zones at greater distance and the halides will start depositing on to the shell. On the other hand if the distance is smaller, tungsten halides will not form due to high temperature when halogen will remain as free gas. So for every wattage, voltage, filament cross-section, the proper diameter of the linear tube is to be ascertained before manufacturing a halogen lamp.

Just due to regeneration cycle itself, the halogen lamp does not get a longer life. It has been seen practically that the tungsten molecules released by the tungsten halide, try to settle on the colder regions of the burning filament. From our experience we know that there is temperature gradient even on the length of the filament. The filament being placed longitudinally, it needs support. The contact places of these support carry away heat directly to the shell wall. So, more deposits will be seen near to these supports and the tall ends of the filament. Actually these becomes 'cold spots' of the filament. In between the supports the filaments is hotter, at the centre of the total filament, the area is hotter, so these will be turned into 'hot spots' and these spots will loose filament material at a faster rate. In a halogen lamp the tungsten migrates from 'hot spots' to 'cold spots'. Generally a halogen lamp does not 'conk out' during burning but the failure of the lamp occurs during switching on when great resistance is felt where the filament got thinner due to hot spots.

A great credit has been always attributed to halogen regeneration for the higher efficiencies, and longer life of halogen lamps. In fact the efficiency of any light depends on the 'heat retentive' quality of the apparatus. Undoubtedly the halogen saves the filament but the size and shape of the halogen lamp are the greatest factors where the maximum utilisation of heat could be made from a known source of energy.

The optimum diameter of a 150 watt GLS Lamp is 171.5 mm and this has been reached by taking into account the lumen and life efficiency expected from such a lamp. The length is 345 mm and if this is made shorter or the diameter is reduced, the lamp will get blackened much earlier—on the other hand if the diameter or length is increased, the lumen value will fall drastically due to convection of heat. Whereas a 1500 watt halogen, lamp will be only 10 mm in diameter and 300 mm in length. In fact the halogen lamp has only 1% volume of a GLS lamp.

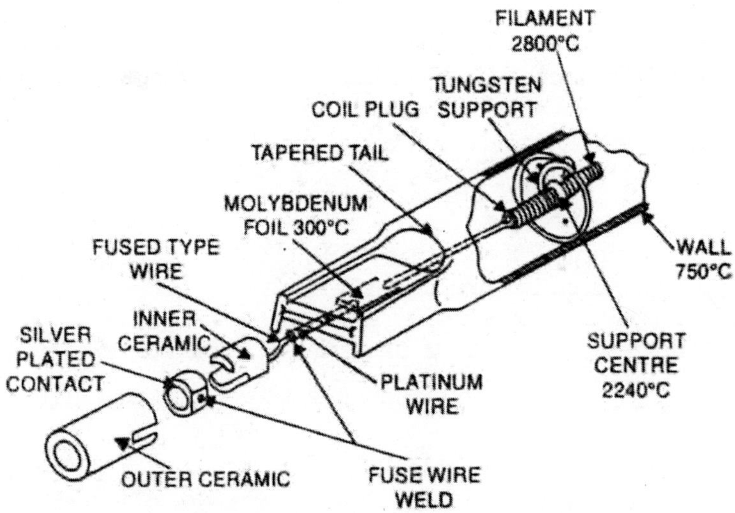

Fig. 8.2. Halogen lamp assembly.

Design Aspects of Tungsten Iodine Lamps

Till now the best material for the filament is tungsten as it has favourable radiation characteristics of a black body as light source. Tungsten has a positive resistance quality which means that it's resistance increases with its temperature. Also tungsten radiate energy over a continuous spectrum from infrared to ultraviolet region. It gives maximum visible light when it reaches its melting point. To avoid the spreading of the conduction and concentrating the heat to a minimum area, the tungsten filament have been coil coiled.

In vacuum condition, the heat quantas from the filament travels unobstructed to the glass shell carrying tungsten molecules. The evaporation rate of tungsten in coiled lamp is faster in vacuum condition. So, inert gas is used to provide obstruction to the disintegration of tungsten molecules. The velocity of the molecules of a gas at a given temperature is inversely proportional to the square root of their atomic weight. So, the heavier the inert gas, better performance of the coiled coil filament can be achieved.

In vacuum condition the tungsten molecules actually 'fly off' from the filament as heat increases due to resistance to the electrical pressure.

In gas filled condition, the tendency of the tungsten molecules to "fly off" prevails but are very much subduded due to the pressure of inert gas molecules.

It has been found by Langmuir that the atmosphere surrounding the filament is very very stable. So, the gas covering known as 'Langmuir sheath' passes the heat by conduction only through the molecules of inert gas around it.

In the above illustration the glass shell is shown of tubular shape to understand the main designing factor of a halogen lamp. The distance between the filament and the glass wall and the gas present around the filament produce the special properties of a halogen lamp.

It has been seen that the Langmuir sheath must contain some inert gas of heavier molecules. The molecular weight of nitrogen is 28.0 and Argon is 39.8. It has been seen experimentally that a pure Argon filled lamp has 3 times more life than a nitrogen filled lamp. Krypton is a very expensive inert gas which costs about 2000 times more than nitrogen. A crypton filled lamp gives seven times more life than a nitrogen lamp. So the best economical answer came to be the use of argon gas to provide the Langmuir conception. In linear halogen lamp as the distance between the two electrodes are very far, which makes it impossible for short circuiting between the ends. So ultra pure argon is used to get better efficiency.

The result of calculations of the tungsten transport rate as a function of the lamp size at a constant inert gas pressure are shown in Fig. 8.3.

Selection of Optimum Diameter of Quartz Tube for Halogen Lamp

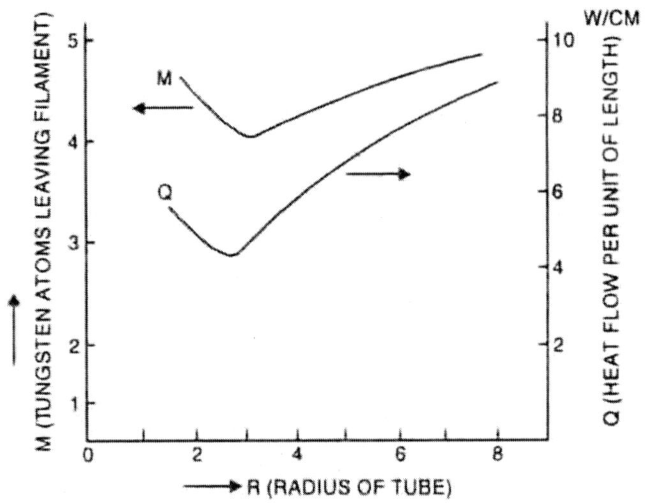

Fig. 8.3. At an argon pressure of 10 (atoms) with a tungsten wire diameter of 1 mm and a wire temperature of 3200 K, the curve shows the optimum diameter of the quartz tube as R ≈ 3 mm.
(*Courtesy* : Phillips Tech. Rev. (35–299–1975) Holland).

Glass Shell. This is the most important item in halogen lamp. A halogen lamp operates at a temperature nearing 3000°C which is more than 2½ times greater than the melting point of silica. The inner surface of the halogen lamp must maintain a temperature of 300°C. At the same time protection is to be provided as halogen is a very corrosive gas which also attacks glass. The glass should have minimum coefficient of expansion to avoid leakage through the electrode entrance holes.

Considering all the above points, it has been seen that the conventional soda glass or lead glass tubings would not suit for the covering of halogen lamp. So the glasses used presently are either vitreous silicon or corning vicor.

(*a*) *Vitreous Silicon.* This is pure silicon dioxide having very negligible amount of impurities (maximum 1%) and known as "quartz glass" because of its very clear transparent nature. It is manufactured by gas fired or vacuum process.

To provide resistance to chemical corrosiveness, every effort is made to reduce the hydroxyl content. While drawing, this glass also must have no air lines on it. Generally high quality quartz crystal (called Brazili n quartz) is acid washed, crushed and melted in a carbon mould by induction heation under vacuum condition. The silicon thus obtained is cooled and again placed in the furnace in a reducing atmosphere. From this furnace, the molten material is drawn through the orifice and mandrel of desired dimensions. Though very expensive, this method produces the best of quartz tubing.

Quartz glass also are being made in controlled atmosphere furnace where the crystals are melted in an atmosphere of 90% nitrogen and 10% hydrogen. In another method, the powdered crystals are fed through a flame on to a rotating mandrel which is called the flame fusion method. In both these processes, the hydroxyl content remain high and also the tubings have more air lines.

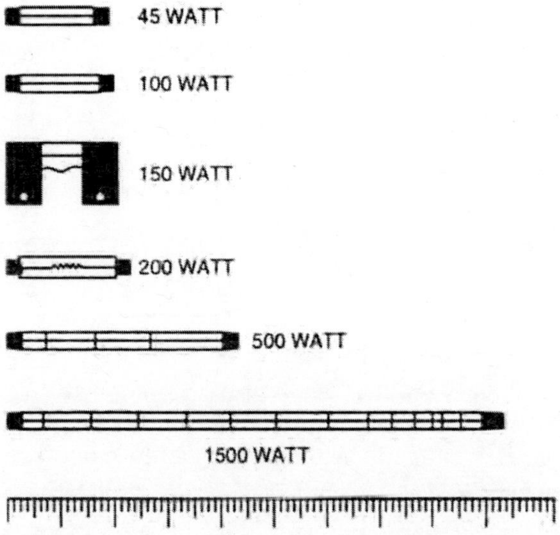

Fig. 8.4. Sizes of different quartz iodine lamps.
(Courtesy : General Electrical Co. Cleveland, U.S.A.)

Corning Glass. Corning brand is the trade name of very high quality borosilicate glass which has good resistance to mechanical and thermal shock and chemical attack. In future this glass may be manufactured in India with technical collaboration of corning glass works, U.S.A. and Borosil Glass Works Ltd., Bombay-14. Presently marketed corning products are made to composition code no. 7740. For making halogen lamps, we need Corning's Vycor glass made to composition code no. 1 corning 7913. Many enthusiastics have tried in vain to produce halogen lamp with corning 7740 without knowing the constituents of these two types of glasses made by the same company.

Composition by weight	*Corning 7740 generally known and used as corning glass*	*Corning 7913 used for halogen lamp known as Vycor*	
SiO_2	81	96 }	.... It makes 99% pure boro-
B_2O_3	13	3 }	silicate glass. Rest 1%
Na_2O	4		being negligible chemical
Al_2O_3	2		and impurities.

For making Vycor corning glasses, the normal borosilicate glass is leached in acid. The acid eats away all the constituent excepting the silica and boric oxide. This action leaves millions of holes in the glass. The glass is then placed under high temperature which consolidate the porous structure and the material shrinks to 35% of it's volume. The product thus produced gives Vycor corning 7913 glass the suitability for making halogen tube shell.

The following comparison of the properties of both these corning glasses shows the suitability for selecting Vycor corning glass for halogen lamp tubing :

	Corning 7740	*Corning Vycor 7913*	*Vitreous silica*
1. Co-efficient of linear expansion at °C × 10^{-7}	32.5°C	8	5
2. Annealing point °C...565C	565°C	1020°C	1150°C
3. Softening point °C.........	820°C	1530°C	1600°C

For halogen lamp the surface of the shell must have minimum temperature of 350°C. Naturally with 800 to 1000 watt lamps, the shell tube temperature is very high and would cross the annealing point and softening point of corning 7740. Also the Vycor quality has so low co-efficient of expansion that the sealing ends would maintain leak proofness 4 times more efficiently than corning 7740.

The best suitable glass for manufacturing halogen lamp is vitreous silica material but the glass industry in India as yet does not have the requisite facilities to produce the same. This is also expensive in comparison to corning 7913 glass.

Unless Vycor 7913 or vitreous silica glass tube is used, all attempts to manufacture halogen would prove futile under conventional process.

Sealing or Pinching Problems. In a linear type halogen lamp, we need two electrodes (lead-in-wires as incandescent lamp) at both ends. The tungsten filament of a halogen lamp works at an optimum temperature of 2800°C. So, to hold the two tails of the filament, we need a material which would not soften at that temperature and also have a minimum coefficient of expansion. In a normal incandescent lamp, nickel iron electrodes are used and are coated with sodium borate at the place of 'pinching or sealing'. The problem of sealing in halogen lamp becomes very critical because of high working temperature. Here the electrode used is a thin molybdenum sheet of about .025" thickness. At one end the tungsten filament is welded and the full sheet is then pinched between the softened Vycor or vitreous glass. The electrodes—the two molybdenum foils at the end of the tungsten filament heat up to 1300°C and being metal, they have good amount of expansion at this temperature. On the other hand, the vycor of vitreous silica tube has very very low coefficient of expansion. The thin molybdenum foil which actually had crinked and lodged itself inside the vitreous tube material is under great stress. Here the vitreous material should have such cohesive quality at 1300°C to withstand the stress produced by the molybdenum foil and at the same time the molybdenum foil also should not crack or fracture under the pressure from vitreous material and continuous passing of heat and current through it. Actually the ductility of molybdenum helps to retain all the characteristics necessary for it's suitability to halogen lamp. At the outer end of the molybdenum plates, a molybdenum wire is welded for electrical connection. The exposed moly-wire is given a platinum coating for avoiding oxidation of the moly wire at temperatures above 350°C. The exploded view of the end connections are given in Fig. 8.2.

Filaments. To select tungsten material suitable for working in a halogen lamp, very good care has to be taken. The purity of tungsten is the most important factor. In a halogen lamp no other material except tungsten and molybdenum can be introduced. Specially, for supports, either tungsten, molybdenum or ceramic is to be used otherwise any trace of iron or nickel will form hallides themselves and stop regenerative cycle of tungsten halides.

The tungsten coil is made specially for halogen lamp. Actually these are more closely wound to give more heat and light per unit area and the coil should not sag at any time during its burning life. In halogen lamp the filament also reaches temperature upto 2800°C. To maintain rigidity of the coil at this temperature the tungsten wire is specially treated in hydrogen atmosphere to remove the fibreous structure while drawing through dies to a crystalline structure. At high temperature, these crystals of tungsten with the body of the lattices having better bonding qualities do not slip from their own positions and avoid sagging of the coil.

The low wattage halogen lamps have single coil filaments whereas the high wattage lamps use coiled coil filaments.

The optimum filament temperature for working of a halogen lamp is 2700°C to 2800°C. The halogen lamps have got to be made inside a tube so that the gap between the outer surface of the filament and inner surface of the tube lie at its shortest endurable distance for the vitreous material to withstand the heat produced. The tungsten halide regeneration functions only within the stipulated limits. Being of linear construction, the full length of the filament has to be used without any bend. This necessitates placing of supports to the filament. Generally for long filaments, a support is put at every centimetre length. For support material, molybdenum may be used in lamps with lesser life as it has lower melting point and evaporates itself on continuous use. Generally 0.2 to 0.3 mm tungsten wire is used as support wire. These radial supports have got to be made very accurately and the inner circle centre of all the support must lie in one line. The following illustration shows the mounting of the tungsten support wire.

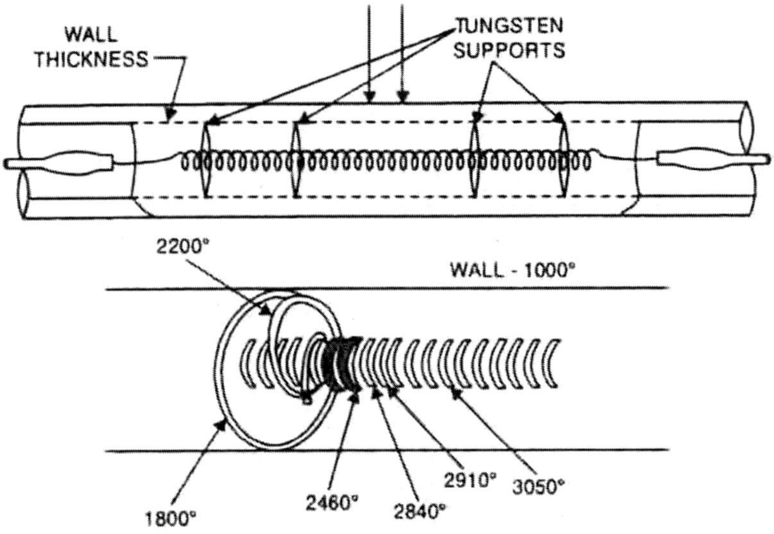

Fig. 8.5. The temperature distribution (°K) produced by a coiled support and practical placement of four coiled tungsten supports (above) which extend to the quartz wall.
(*Courtesy* : Mr. R.J. Campbell, GEC, Cleveland, U.S.A.)

TYPES OF HALOGEN LAMPS

1. **Tungsten-Iodine Lamp.** Out of the five members of the halogen group, iodine received the first preference for application in the halogen lamps. The very first advantage being its capacity to start the transport cycle of tungsten halide at 1000°C which was easier to be achieved by the available raw material at the time of development of the lamp.

Iodine is a solid at room temperature and melts at 114°C. It's boiling point is 184°C. It is easily available and also handling is much easier. For using iodine only care is to be taken to make it perfectly dry as it catches moisture very easily. For using iodine, the same type of GLS lamp gas filling tower material is to be used. The iodine is dried by sublimation in vacuum over phosphorus pentoxide. For all halogen lamp making, rubber tubes are to be avoided as they attack these vigorously.

The iodine reaction takes place at 1000°C as iodine (I) + 2 tungsten (wolfernite—W) = WI_2 (tungsten halide). But a successful transport cycle starts functioning at 1700°C of the filament temperature and about 350°C of the shell temperature.

Iodine is dark brown in colour and gives off pink vapour. The quantity of iodine to be introduced is dependent on many factors. If the quantity is less, full generative cycle will not operate and tungsten deposit on tube on the wall will be seen. If the iodine quantity is more, the tube wall will become pink to dark brown in colour restricting high percentage of light. The quantity of iodine to be introduced is dependend on the wattage of the lamp, the inside area, the distance between the filament and the glass wall and the type of inert gas filling at cold pressure. Naturally for the best result, the optimum quantity is to be determined with every single type of lamp keeping all the materials and design factors constant and varying the amount of iodine input. Through 5–6 experiments, a good result can be obtained. Even the designers of halogen lamps agree that a minimum of 5% light efficiency will be lost due to use of iodine inside a lamp. For selecting a halogen lamp it can be said that a halogen lamp filled with a little excess of iodine will give slightly greenish spectrum of light.

Presently iodine is used in long-life lamps.

2. **Metal Halide Lamps.** Here mention also has to be made of METAL HALIDE LAMPS which may sometimes create confusion with conventional incandescent halogen lamps.

(*See Photo 8.1*)

Metal halide lamp is a type of discharge lamp and is quite different in operating principle with incandescent lamp. This is introduced here to inform the readers as these lamps also use the regenerative principle of the halogen elements. Metal halide lamps are in a very active state of development as lamp size is made even shorter and the use of filament is avoided. As the breakage of the filament is the main cause of failure of halogen lamps, the metal halide lamps may become the lamp of the future.

The metal halide lamps have two electroded placed at a distance between 9 mm to 14 mm depending on the lamp type. The electrodes are of pure tungsten and thoriated to avoid oxidation in the presence of halogen. At present there are seven iodides (halogen), as sodium iodide, thorium iodide, scandium iodide, thallium iodide, indium iodide, dysporsium-iodide and gallium iodide used either singly or in combination with 3-4 of them to get better results in metal halide lamps.

The compact type metal halide lamps are very popular for high speed photography, search lights, sports stadium and optical instruments. These lamps are available from 750 W to 10 kW range.

Metal halide lamps are also slowly replacing the mercury vapour lamps. These lamps are made with elliptical outer envelope and available from 400 W to 1500 W.

In metal halide lamps, at a certain temperature, the metal iodides dissociate as iodine and the metal. This also produces a little pressure inside the lamp. All the atoms of metal, tungsten and iodides are in constant movement—either separating from one another or forming iodides depending on the temperature zone where they are placed. At the same time the combination of more than one halide changes the colour spectrum of the lamp—in the case of coloured light source only thallium iodide serves the purpose.

The only disadvantage being that these lamps need a starting device as choke ballast like mercury vapour lamp.

3. **Tungsten Bromine Lamps.** For all the halogens, the regenerative cycle functions in the same manner. Iodine and bromine cycles also start working in the same way when introduced in halogen lamp of similar shape and size. But bromine-tungsten transport cycle starts working at 1500°C. So, the lamp must reach this minimum filament temperature otherwise the bromine shall start acting directly upon the colder region of the tungsten filament. Bromine is much more reactive than iodine and the free bromine would eat out the filament.

To overcome this difficulty, HBr_2 (Hydrogen Bromide), CH_2Br_2, CH_3Br_3 is presently used instead of bromine in pure state. So, when due to rise in temperature beyond 1500°C.

Bromine is rebased, it gets thrown off by the tungsten molecules to react with it. After prolonged running, the hydrogen of these compounds, slowly leaks through the various joints of the halogen lamp and the bromine becomes free to react on the cold tungsten filament. For this reason, higher wattage value and higher lumen content halogen lamps are made with bromine compounds, where life expectancy is less than 1000 hours. The amount of bromine gas to be introduced should be very exact as any excess will attack the cooler part of the tungsten filament.

Advantages of Bromine Lamps

1. The filament can be operated at a higher temperature of 2800°C giving more lumen per watt.

2. Bromine compounds are introduced in gaseous form and makes it easier for operation.

3. The filling of bromine gas does not create any problem as there is no light absorption. So, it gives always 5% more light than iodine lamps.

4. The range of regenerative cycle is wider in tungsten-bromine lamp where the bulb shell temperature operates between 200°C to 1100°C.

The main disadvantage of high filament temperature lamp is it's thinning out of the filament at the hot spots. In halogen compounds, the traces of oxygen, hydrogen, water and minor impurities can never be fully eradicated. This forms tungsten oxide with the 'thrown out' molecules of tungsten and settle down at the colder region of the filament. Being oxide, it burns out the filament pre-maturedly.

The presence of water in a halogen lamp causes great retardation in it's development. Even after taking utmost care, some molecules are always present in the lamp. It is not only due to bromine compound but water vapours are found even when the lamps are filled with ultra pure argon at 10 atmospheric pressure. Though ultra pure, there will always be some residual traces of water vapour in the inert gases.

This oxygen and water vapour difficulty has been overcome by using special getter compound known as bromophosphonitrate ($PNBr_2$). As described earlier just like GLS lamp gettering, a little quantity of $PNBr_2$ used as getter instead of red phosphorus getter. $PNBr_2$ is a polymer which is chemically inert and does not react with either glass or tungsten also this is non-poisonous in nature and very safe for handling. This comes into operation at the time of ageing when the phosphorus first gets burnt and neutralises the oxygen with other impurities. The bromine sticks to the tungsten filament till the transporting temperature is reached and the regenerative cycle starts operating. For increasing the bromine content of the compound, instead of increasing the concentration of the $PNBr_2$, it has been found to be ideal to use trimeric bromophosphonitrate $(PNBr_2)_3$ or tetrameric bromophosphonitrate $(PNBr_2)_4$.

By this method of introduction of halogen through gettering has greatly eased the production process.

The tungsten bromine lamps are manufactured for projector type flood light lamps with unsupported filaments. The unsupported filament does not have 'cold spots' and the deposition of tungsten is fairly distributed.

4. **Tungsten Chloride Lamps.** The main advantage of using chlorine in a lamp is it's easy accessibility and it's introduction into inside the lamp in gaseous state. The production

procedure can be simplified as in conventional GLS lamp manufacturing. With heavier molecular weight, the performance of the transport cycles is better than iodine and bromine. The disadvantages of using chlorine are also manifold—very poisonous and injurious for inhalation, handling difficulties and corrosive to normally used lamp manufacturing accessories. Chlorine lamps are still in the research stage and it also holds a hopeful commercial usage.

5. **Tungsten—Fluoride Lamps.** Of all the halogens, fluorine is expected to outclass the others theoretically of this has the properties to activate a near perfect regeneration cycle. We are now aware that the failure of a filament occurs due to 'hot spot' formations and various technical factors which shall always remain uncontrollable. The introduction of halogen has opened a new venue to induce automatic repairs to this 'hot spot' problem. Iodine, bromine and chlorine due to their individual properties dissociate at different temperature from the tungsten and start the regenerative cycle or 'internal repairs' to the worn out filament. The disadvantage of these halogens are that they always try to accumulate at the colder region of the filament.

The main advantages of fluorine are that this would dissociate from the filament above 3000°C. So, at the point of the filaments, where the hot spots occur and go beyond 3000°C—the fluorine will dissociate and start the regenerative cycle with tungsten. Test have shown that the tungsten—fluoride, fluoride—tungsten transport cycles are localised at these hot spots only and do not spread out on to the complete area of the lamp. The tungsten thrown off at the hot spot combine at the same spot as tungsten fluoride and at the same instant decompose as two elements of tungsten and fluorine and the tungsten settle back again on the hot spot from where it was thrown off releasing the fluorine to start another cycle. Theoretically this speaks of a infinite filament life. Tungsten melts at 3320°C and efforts are being made to manufacture filaments with resistance to produce temperature nearing the melting point. We know that the higher the temperature concentration, the better lumens value per watt can be achieved. Till now increase in temperature was restricted due to heavy dissociation rate of tungsten beyond 2600°C. Now with introduction of halogens and particularly with the prospects of fluorine, very fascinating results have been achieved. Though there are innumerable technical difficulties to achieve the tungsten—fluorine regenerative cycle, it has the greatest promise for the future.

Halogen Lamp Manufacturing Process

The higher wattage halogens are manufactured by hand process with the aid of semi-automatic machines. The miniature halogen lamps used in automobiles are made in large numbers in automatic machines.

(*See Photo 8.2*)

The construction details of a 240 V/1000 less W Halogen Lamp.

1. Take a corning Vycor or it's equivalent vitreous glass tube of 100 mm length and 10 mm outer diameter.

All work on this material must be carried out with oxygen supported gas flame.

2. Close one end of the tube, heat the centre of the tube with a very sharply pointed flame, blow from the open end and make a hole. Join a 3 mm to 4 mm outer dia vitreous glass tube of 50 mm length over this hole. This will be used for exhausting, gas filling and final sealing.

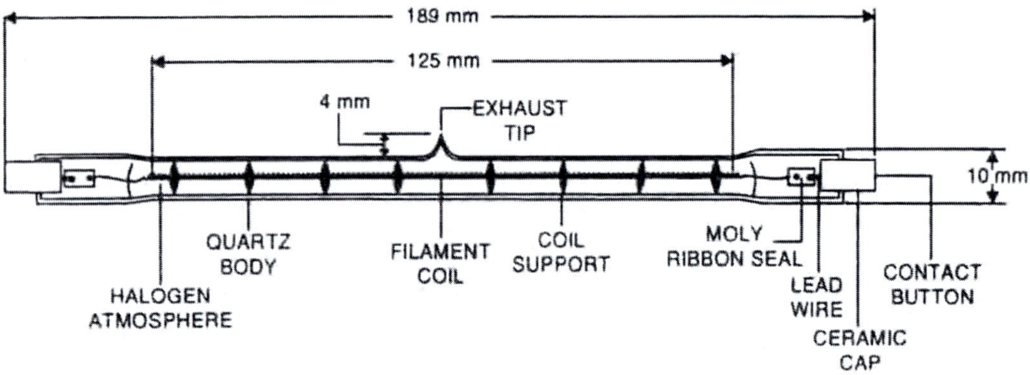

Fig. 8.6. Dimension details of a quartz halogen lamp.
(*Courtesy* : Litex Electricals Pvt. Ltd., Pune (India))

3. Make a solution of red phosphorus with cryolite in solution of methyl acetate and chloroform spirit. Mix a little quantity of tetrameric bromophosphonitrate $(PNBr_2)_4$.

Now dip the selected filament, the four support rings and the moly plate in the solution.

4. Select the proper filament. Make four rings with 140 micron thick tungsten wire. The inner circle of the ring should just touch the ring and the outer circle should be just a little larger than the inner diameter of the tube. A template is to be made tc make these ring. All these rings should be of same size. While inside the tube, the centre of all these rings should lie in a straight line and also should coincide with the centre axis of the tube.

Every thing in a halogen lamp depends on this mounting. If the filament does not rest at the centre axis of the tube, it will cause hot and cold spot, uneven burning and uneven regeneration cycle.

The diameter of the inner hole of the ring should be such that the filament could just pass through it without difficulty.

Place the four rings at the equal distance from each other inside the tube.

Take two thin molybdenum sheets preferably cadmium plated 8 mm length, 4 mm breadth of 0.2 mm thickness. Have a 0.5 mm thick tungsten wire welded at one end.

Insert the filament through the ring on one end of the tube and on the other end have a tag attached. When the untagged end comes out of the tube, spot weld end of the molybdenum plate to it. Now take out the free end of the filament through the other side of the tube by holding the tungsten-wire end of the molybdenum plate. Thus mounting is completed.

Place the mounted tube on to a zig. Arrange to keep the molybdenum plate at the middle of the tube by fixing the tungsten electrodes on to a holder.

5. Heat up both the ends and pinch them at the molybdenum plates making a seal-proof joint.

6. Evacuate the tube through the exhausting port at the centre. Use little heat. Do not overheat which might evaporate some of the halogen getter.

Photo 8.1. A high pressure metal iodide lamp (HPI). The lamp is made with a tubular clear envelope. (Courtesy : Philips Tech. Rev. (35-11/12) Holland.)

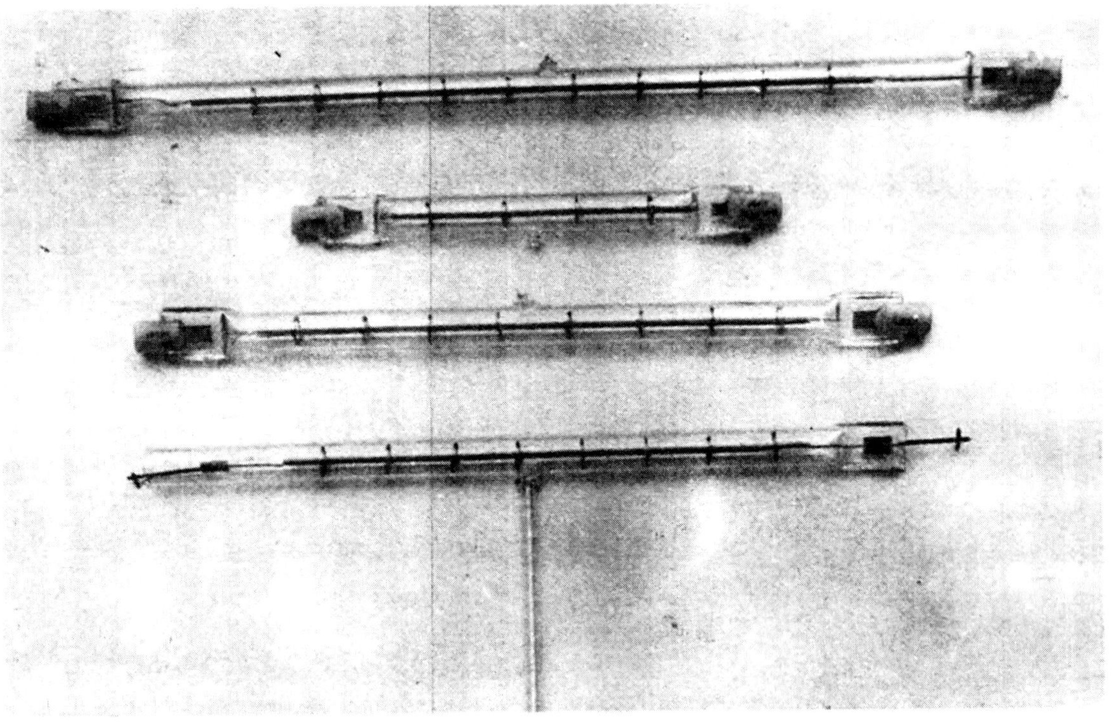

Photo 8.2. Halogen flood light lamp.
The bottom piece showing the mounting, pinching and exhausting procedure.
(Courtesy : Litex Electricals (Pvt.) Ltd., Pune, India.)

Photo 8.3. Automatic glass tube crack-off and sorting machine for halogen lamp manufacturing.

(Courtesy : Badalex-a Div. of Sale Tinley Technology—England.)

Photo 8.4. Automatic mount mill-halogen lamp.
(Courtesy : Badalex-a Div. of Sale Tinley Technology—England.)

Photo 8.5. Automatic exhaust machine-halogen lamp
(Courtesy : Badalex-a Div. of Sale Tinley Technology—England.)

Photo 8.6. Halogen lamp focussing machine.
(Courtesy : Badalex-a Div. of Sale Tinley Technology—England.)

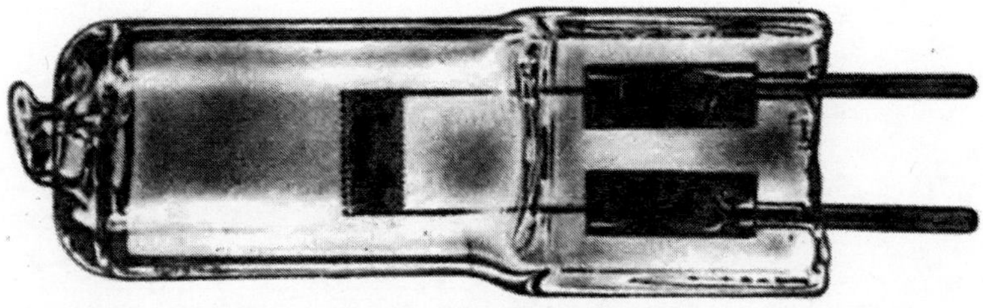

Photo 8.7. Halogen projector, lamp.
(Courtesy : Litex Electricals Pvt. Ltd., Pune, India.)

7. Fill in the tube with ultra-pure argon in cold condition by keeping the quartz tube dipped inside liquid nitrogen. The pressure inside the tube should be around 10 atmosphere.

8. Put the lamp to ageing, starting from 50 V and raising to 200 volts in quick succession raise slowly from 200 volts to 250 volts carryout ageing for 10 mts at 250 volts. For ageing, care must be taken to shield the lamps behind glass screen otherwise there will be harmful effects of ultraviolet radiation to eyes and skin.

9. Do not touch the body of the glass tube with fingers. Hold the lamp at the ends only.

10. Fix the ceramic holders at the ends.

The lamp is now ready for use.

Special Note. If a long life lamp is to be made, after evacuating it, a measuring apparatus for pure and oxygen free iodine gas is to be made (or obtaining oxygen free iodine gas the process is described earlier) and introduced inside the lamp. In that case special Bromine getter should not be used.

For short life and small dimension lamp, Bromine based special halogen getter will serve the purpose.

Advantages of Halogen Lamp. The slogan for halogen lamp is "more light longer life". Till now the best material for incandescent filament has been found to be tungsten. The main disadvantage in the use of tungsten has been it's properties to disintegrate with the rise in temperature. Theoretically it has been seen that the maximum light that tungsten could produce is 53 lumen per watt at molten state at 3410°C temperature. Practically it would not be possible to hold tungsten at molten state and receive the benefit of it's fullest efficiency. Presently a maximum of 40 lumen per watt efficiency has been achieved with a very short life of 3 to 4 minutes. As heating can be concentrated with minimum losses due to gas filling in conventional lamps, the halogen lamp with its very compact design and regenerative cycle, have raised the efficiency of incandescent lamp upto 50%. With different type of halogen members, the temperature in a lamp can be raised between 2600°C to 3200°C with efficiency from 15 lumen/watt to 35 lumen/watt.

Volume. The volume of a halogen lamp is nearly 1% than of conventional lamp. This has helped in heat concentration through increasing temperature of filament. The size and shape of the holders now are reduced to minimal size in comparison to conventional lamps.

Application of Langmuir Sheath. In a perfect Langmuir model, the low density gas layer should not produce any convecting air current. The heat of the filament must flow out by mere gas conduction process only. But in an conventional lamp this is not possible as the distance of the shell envelope has to be kept far away from the filament. So, the low pressure inert gas starts convection current from filament to shell which is undesirable and makes it an inefficient lamp. In halogen lamp, the shell has got to be nearer to the filament to maintain the halogen regenerative cycle. The inert gas also is put at higher pressure varying between 2 to 10 atmospheric which practically retards evaporation of tungsten. In fact, there is no convection of heat by inert gas and it makes fullest use of Langmuir sheath model giving higher lumen value of the filament.

Application of Halogen Lamps

Being linear in shape, the application of halogen lamp was restricted to horizontal usage in the lamp holder. The halogen vapour would accumulate at the upper end of the halogen tube when placed at an angle more than 4° from the horizontal plane. This would naturally affect the tungsten-halogen regeneration cycle. This difficulty has been overcome since using inert gas at a higher pressure and using small amount of xenon gas. Presently halogen lamps are used in linear and 'U' shape. Even, small luminaires as the size of a miniature lamp has been made for lighting the runways.

For automobiles, three types of halogen lamps have been standardised. These lamps are now used as spot light, for light and head lights.

H (Halogen) 1–12 V–55 watt. A single axially placed filament is used in this lamp.

H_3–12–55 watt. This also uses a single filament but usually placed transversely with the horizontal axis.

Both the above lamps are prefocussed type with accurately aligned filament is relation to prefocus ring. Due to low voltage, high wattage very frequent switchings, the normal auto-lamps have a low life rating. Though started with iodine filling, these lamps now mostly use bromine gas because of a easier production procedure. Both these H1 and H3 types are manufactured by automatic machines.

The other standardised halogen lamp is H4 type. This is the type of head lamp found in general use. This has two filaments—one light up for driving and the other is used for dipping (known as meeting beam). In manufacturing this lamp, difficulties were faced as the evaporated bromine from the burning filament would corrode the colder filament and the life would be shortened namely for dipper filament. To overcome this difficulty, the meeting beam filament was sufficiently isolated from the main head light filament. At this place a very negligible quantity of free bromine is available to corrode the cold meeting beam filament.

PRODUCTION OF HARD GLASS H4 HALOGEN AUTO LAMP

Presently H4 type halogen auto head lamps are being manufactured with 'hard glass' instead of 'quartz glass'. In the following pages a typical set up of the machineries and the production procedure of hard glass H4 halogen auto head lamp has been described. All these machineries developed by M/s Badalex, England are described here to illustrate practically the operations being followed to manufacture H4 hard glass lamps and should not be taken as the only available means.

Manufacturing H4 Type Halogen Auto Head Lamp—Automatic Process

With further development in halogen lamp technology, the H4 type auto head lamps are now being made with Hard glass instead of quartz glass. It has been described earlier about the essential conditions and requirements necessary to continue re-transport cycle inside a halogen lamp. One of the most important need is the very accurate positioning of the filament which would not be possible if the internal diameter and the wall thickness of the glass are not maintained within prescribed limits of tolerances. So the hard glass used should be of machine blown type with very accurate dimensions. The length of the filaments being standardised, great accuracy also has to be maintained for the length of the H4 auto head lamps. Because these operations demand great accuracy, all these glass cutting, a separating and sealing operations are carried out in finally adjusted automatic machines. More attention is given to the accuracy in the production of auto head lamps because H4 is a prefocus lamp and at the focusing stage of production all the inaccurately made lamps get rejected.

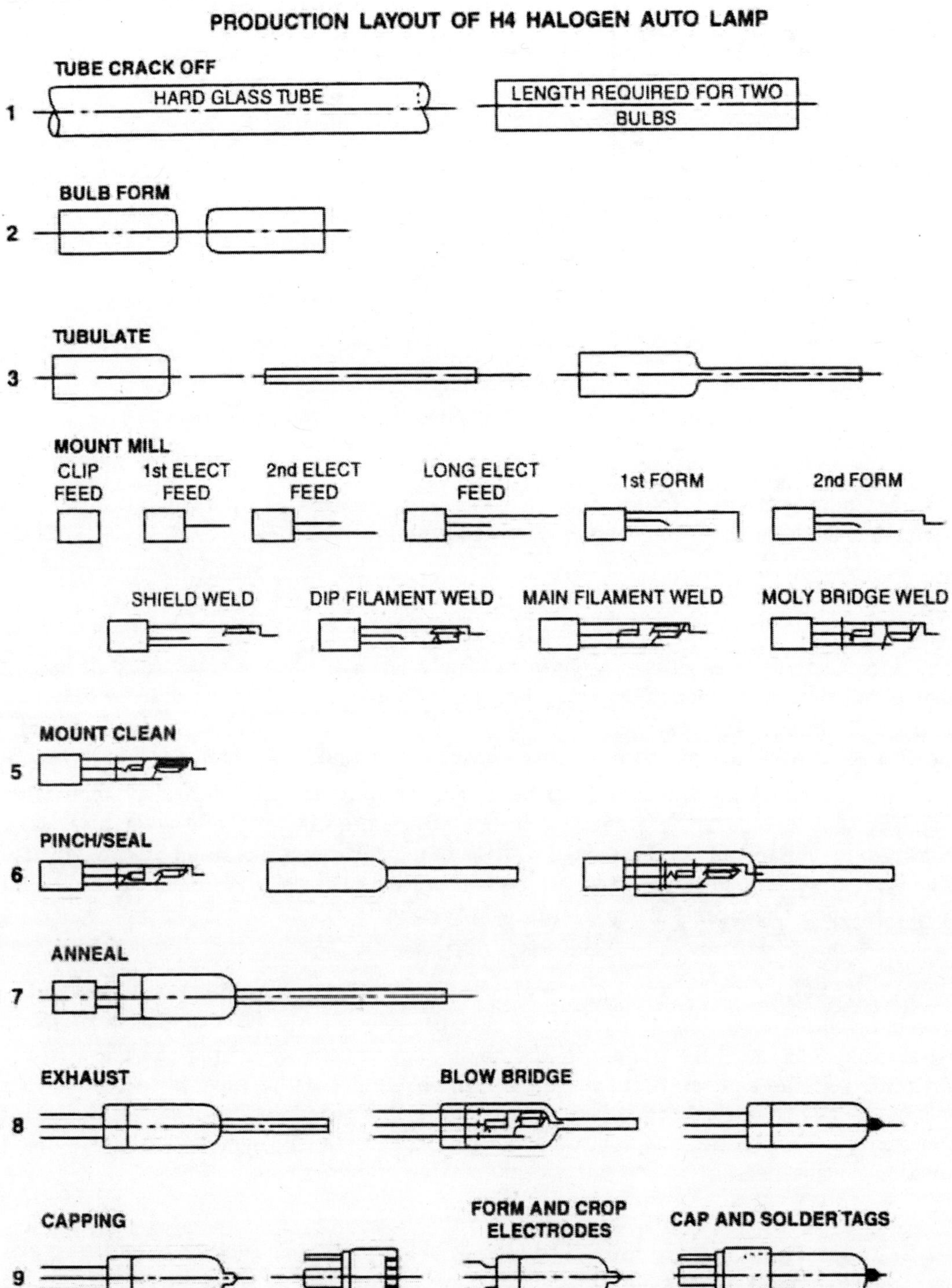
PRODUCTION LAYOUT OF H4 HALOGEN AUTO LAMP
TUBE CRACK OFF
1
HARD GLASS TUBE
LENGTH REQUIRED FOR TWO BULBS
BULB FORM
2
TUBULATE
3
MOUNT MILL
CLIP FEED
1st ELECT FEED
2nd ELECT FEED
LONG ELECT FEED
1st FORM
2nd FORM
SHIELD WELD
DIP FILAMENT WELD
MAIN FILAMENT WELD
MOLY BRIDGE WELD
MOUNT CLEAN
5
PINCH/SEAL
6
ANNEAL
7
EXHAUST
BLOW BRIDGE
8
CAPPING
FORM AND CROP ELECTRODES
CAP AND SOLDER TAGS
9

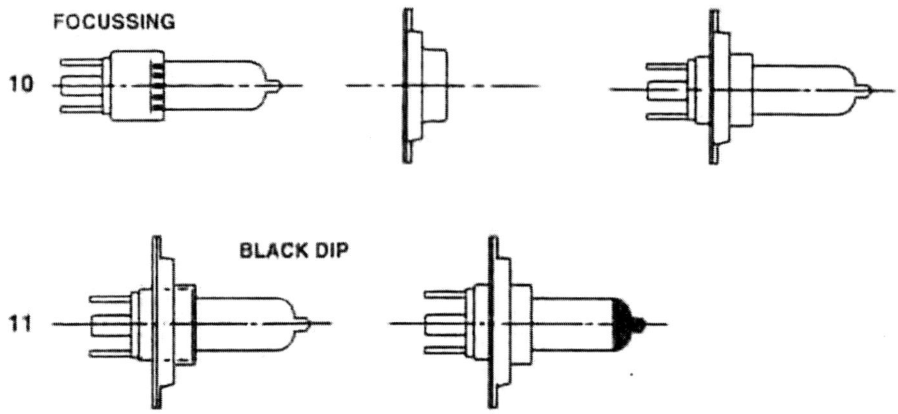

Fig. 8.7

(*Courtesy* : Badaex Sale Tinlay Technology Ltd., England.)

H4 lamps generally follow the three main sequences of operations *i.e.*, 1. Glass preparation 2. Stem mounting and 3. Final assembling.

1. Glass Preparation

Operations

Glass Cutting and Sorting. Glass cutting is carried out by a tube crack-off machine. Tubes upto 1.5 metres in length are manually loaded into a rack. The tubes are automatically fed into the turret of the machine where they are automatically heated, cracked-off and the ends fire-polished. Each of these lengths are meant to make two lamps.

After fire-polishing of the ends, these tubes are automatically fed to a "sonar sorting unit". The tube is indexed to a gauging station where the diameter and wall thickness are measured by means of sonar device and the measurement retained in an electronic measurement system so that at the subsequent stations the tube is unloaded into hoppers depending upon their measuring tolerances. (*See Fig. 8.7*)

Operation

Bulb Blowing. The pre-cut and selected lengths of tubings are fed to a horizontal bulb blowing machine where the bulbs are formed between two pairs of opposite heads. The central portion of the tube is heated, the heads move apart stretching the glass which is then parted at the centre. This leaves a tipped-off length of the tube at each end. At the next station the tipped-off end is heated, a mould closes around it, air is blown through the hollow chuck forming the shape of the bulb in the mould. The mould then opens and the bulb thus formed is sent to the next machine.

Operation

Exhaust Tube Sealing and Tabulating. From the transportation conveyor the bulb is fed to the upper jaw of a multi-head indexing turret. The exhaust tube is fed into the lower jaws. The bulb and the exhaust tube pass through heating process until the two are sealed together and then the hole is pierced. The seal is then annealed and tested for cracks etc.

With this operation, the glass preparation process is completed.

2. Stem Mounting (*See Photo 8.4*)

3. Final Assembling (*See Photo 8.5*) **4. Focusing** (*See Photo 8.6*)

For sequence of operation see production layout diagram of H4 halogen auto lamp in page no. 147 and 148.

Further Usage of Halogen Lamps. For flood lighting and photographic purposes, there are many types of halogen lamps. In fact, halogen lamps have started replacing all the conventional lamps. The flood light lamps upto 500 watt use single coil filament and for higher wattage use coiled coil filaments. (*See Photo 8.7*)

Presently 500 W to 1500 W linear type halogen lamps are extensively used for flood lighting. These lamps are usable with 110 V to 240 Volt supply source and have a rated life of 2000 hours.

For providing continuous beam for movie cameras, a sun-gun 1000 watt halogen unit is used. This gun-gun halogen uses a 'U' shaped tube and has 15 hours running life. The illumination and life compared to conventional flood light shows that one 'sun-gun' 1000 watt halogen lamp produces 7 time more light than 275 watts lamp with reflector having only 4 hours running life.

Another important feature of halogen lamp is that its usage as standard for measuring light intensities in the ultraviolet region. The high frequency of the molecules due to higher filament temperature produce a very large emission in the ultraviolet region. The glass shell envelope of a halogen lamp being of pure vitreous material, the ultra violet emissions radiate from the filament without any change. For this reason, in normal use, the halogen tube has a glass shield to cut down this ultraviolet radiation to harmless condition. This extra radiation property of halogen lamp emitting long ultraviolet wave is used to a greater advantage in spectroradiometry for calibrating measurements of spectral power distribution.

Summary. We have seen now how a simple addition of little halogen has opened a new era in lamp making. With all the four halogens actively contributing to the specific need to the demand of light, the use of halogen gas is going to become common phenomenon as introducing argon-nitrogen inert gas in all incandescent lamps.

Any new introduction takes time for conversion to mass products. The large manufacturing units who spend large sums of money in research and development are sure to take advantage to market their product at a premium during this interim period. But the halogen introduction as a common lamp is taking a very long time though low wattage short life lamps have been mass produced for auto industry. This shows a general aversion of the large manufacturers to offer this benefit in general lighting service. A great deal of changes in the production technique and retooling of plant and machinery will be needed. The small scale manufacturers may start halogen lamp production in a pilot scale as their machineries are semiautomatic type and the product can be checked and developed at every stage.

The redeposition of tungsten by halogen has brought a great breakthrough in lamp technology. A time might come when the life of a tungsten filament lamp will not be a matter of importance at all. The other important breakthrough needed now is to excite the filled in gases even with low voltage so that the great advantage of ionisation takes place causing emittance of continuous spectrum without using transformers and ballast chokes as needed for gas discharge lamps.

So, it means research should carry on without stopping at achievements but expanding it's horizon to a limitless boundary.

9

Manufacturing Defects of Incandescent Lamp and Its Remedies

CONTROLLING RAW MATERIAL LOSSES IN SMALL FACTORIES

In day to day running of the factory, it is but natural the faults to occur even at the most unexpected places. The main criteria is the system to be adopted to reach this information to the proper place so that immediate decision could be taken to eradicate this defect.

The very first necessity is the training to all the workers and supervisors alike about the cost-consciousness of every bit of raw material and the loss of every small part at every point of operation.

It is a common sight to see scores of lead-in-wires, solid rod pieces lying around the stem machine, a good number of filament, destroyed due to bad handling and of the cracked shells and lamps there is hardly any notice taken.

In the chapter on manufacturing, many of the defects are already mentioned but the controls needed to check these defects are being discussed here.

The assembling in lamp making process starts from Stem Machine (please see Flow Chart in Chapter 7). All the raw materials start coming to the operation process from this point.

The main item of a lamp factory is glass and a good part of it is lost in dressings and preliminary work before it is put to main operation process. In almost all the lamp factories the old system of main stores-final product counting process is adopted where it becomes very difficult to point out the faulty operation point.

The author has advised the following system to be adopted in the small scale lamp factories. Here the main-store keeper is asked to operate the Work-in-progress [WIP] Store on the work floor. In lamp factories, the raw materials are issued in bulk and the stores need not be kept open throughout the day in small factories. In large factories a person is engaged to man the WIP store. The main point in this system is that the actual 'raw material after initial work', is the main item for which record is most essential. The glass cutting man, flaring man, shell washing man, molywire cutting man thus know their work are being checked also the stem man, the mounting man and the sealing man know their production are under vigilant eyes and are being recorded. So, in a lamp factory the main store-keeping action should be of the WIP store keeper—who keeps all the records of preliminary losses and Secondary issues. In very large scale factories where more men are working in Stores, it is all the more necessary to engage persons at WIP stage who would be directly responsible to the works manager.

Main Store *(opened when necessary)*

(Issued by counting or weighing)

Count and Accept

WORK-IN-PROGRESS (WIP) STORES

Count and issue

Steam m/c	Inserting	Mounting	Sealing m/c

(All to provide final quantity to WIP for work)

The WIP Stores will show the losses made upto operation no. 4. The WIP card at every machine will show the raw materials received from WIP stores or from previous machine and daily issue to the next machine. So, the production loss at any point can be easily calculated.

PRODUCTION DEFECTS AND REMEDIES

The following gives the general defects which contribute to the production loss and the remedial measures to be taken to overcome them.

Defects	*Remedies*
1. Shell Washing	
(*a*) Hazy shells	Reject sodified shells
(*b*) Contaminated shells	Lamp with contaminated shells give away during ageing. Always wash finally with clean water.
(*c*) Dirty shells and Water marks.	After drying keep shells in covered box.
2. Shell Stamping	
(*a*) Internal stamping augments contamination	Introduce outside stamping system. Check standing materials.
(*b*) Blotched stamping	Use steady hand with sharp lettering. Take blotched-stamped shells for rewarding.
3. Glass Cutting	
1. Undressed wheel	Keep the wheel angle properly dressed with diamond point.
2. Low speed rotation	Use proper motor and pulley

3. Shaky wheel, improperly grouted machine.	Change bearings. Grout machine properly.
4. Loosely held glass-tube causing end cracks.	Instruct worker properly.

4. Flaring

1. Flaring at improper temperature	Show right temperature to the worker
2. Flaring at improper angle	Decide the angle for every type of lamp and train the worker.

5. Indexing (Stem Making)

1. Heavy losses during every start.	Heat up all heads individually before use and check adjustment.
2. Black-milky stem pinch	Adjust oxygen or air properly
3. Improperly melted pinch	Adjust burners
4. Hole-less stem	Use hot air for bursting hole and flaring the neck.
5. Dumet point very near or partially exposed off the pinch.	Adjust stem heads
6. Cracking of good stems	Adjust annealer.

6. Button Making and Pig Tailing

1. Black buttons	Use flame ends
2. Improper button size	Use button dies
3. Loose pig tails and uneven heights.	Insert moly wires when button is properly heated

7. Mounting

1. Dirty atmosphere at the mounting place	Make separate cubicles for mounting
2. Dirty hands of the mounting, worker	Instruct worker strictly of the importance of cleanliness.
3. Stretched filament	Instruct worker
4. Flower not properly sprayed out.	The flower should be made around the central axis of the lamp.

8. Gettering

1. Loss due to over and under gettering	Throwaway old getter every morning. Use fresh getter in right quantity. Use stainless steel ball mill for getter.
2. Lamp reddening	Do not getter filament till the last drop in the bowl. Make a mark for refilling on to the bowl.

9. Sealing

1. Uneven sealing shape	Adjust all the burners at same height.
2. Cracking after sealing	Use hot air for shaping after cutting burner.

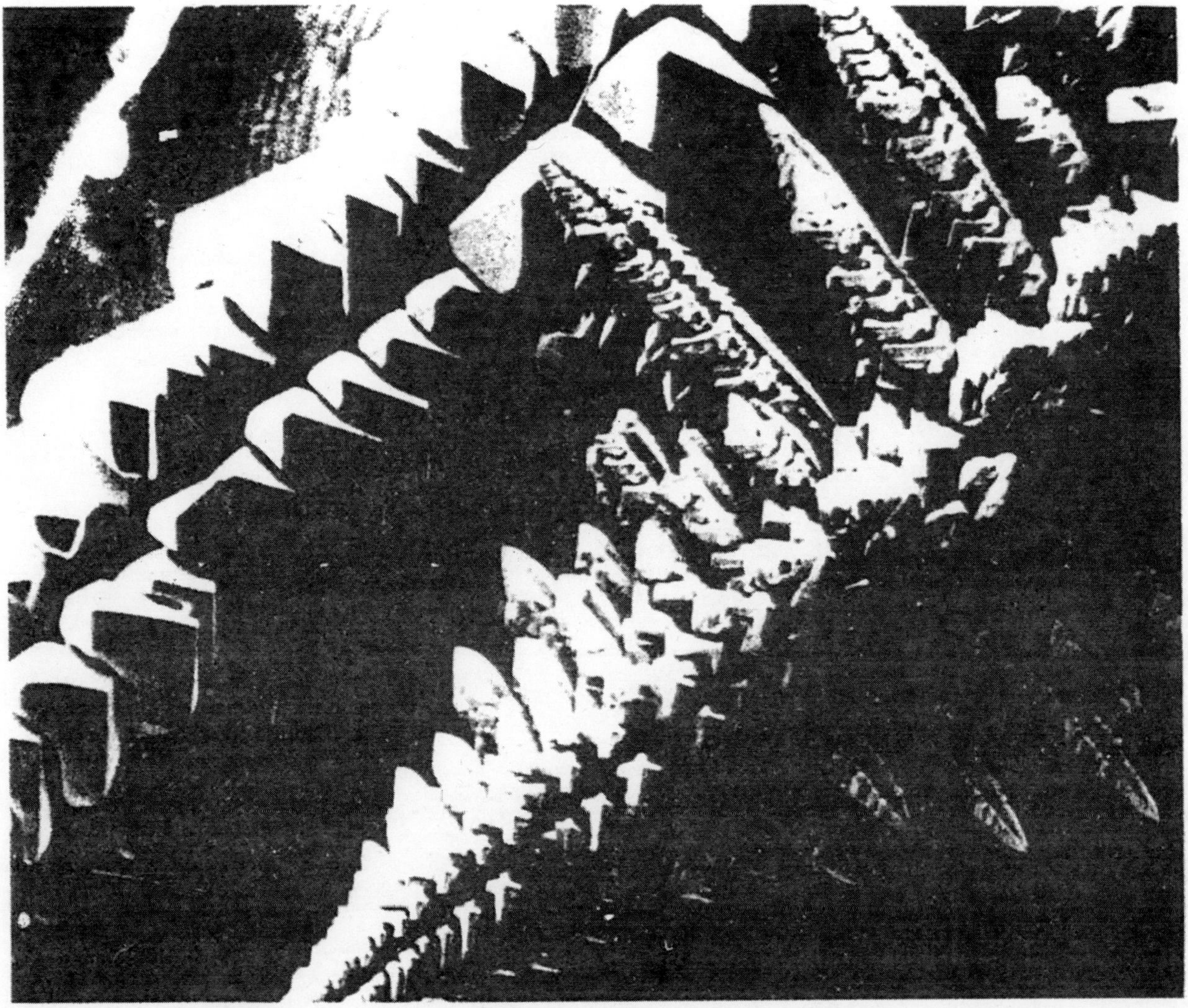

Photo 9.1. The attractive **'spear-head'** shaped dendrites magnified 45 × which are the greatest enemies of lamp makers.
(Courtesy : Philips Tech. Rev. 36 (1976) Holland.)

Defect	Remedy
3. Filament not centred.	Engage skilled person and instruct him about the importance of light centre.
10. Exhausting	
1. Under or over heating of lamps	The life of the lamp depends on proper heating. Use proper heating. Heat till the pinch shows yellow glow when tester is used.
2. Under exhausting	Use meter and tester till the lamp shows proper colour.
3. Improper or extra long tipping	Tipping should not be too long or bent inside causing obstruction.
4. Under filling of gas mixture	Use pre-determined height of mercury column showing 700 to 720 m/m mercury pressure on each lamp.
11. Ageing	
1. Sparking regulator Arm. Rubbing contact stepless regulator is to be used-otherwise good lamps also will give away or weak due to sparking.	Short length filaments as in mini and auto do not require ageing. This is carried on longer filaments.
2. Ageing with more than 10% over voltage	Ageing at more than higher voltage is deterimental to the inner structure of the filament material. It will shorten the lamp life.
3. Short duration ageing	Ageing should be carried out atleast for 5 mts. for each batch so that the filament will get its permanent place in the moly ring.
12. Capping and Soldering	
1. Improper bonding	Use proper capping cement. Test with a torsion tester. The caps against following pressures. BA–22, E–27 = 30 kg cm BA–7, 9, E–10 = 2.5 kg cm BA–15, 20, P–22 = 11.5 kg cm
2. Overheating of lamp neck	Use progressive increase and decrease heating of lamp cap.
3. Faulty soldering	Maintain 2–3 mm height of solder on eyelet.
13. Final Testing, Cleaning and Packing	
1. Accumulation of tested lamps	A tested lamp must be cleaned and packed simultaneously.
2. Improper cleaning	The soldered eyelets must be cleaned properly with wet cloth. The glass body should be cleaned properly with dry-cloth.

14. Quality Control

1. Initial rating, life test, lumen test not carried out.	Test lamps as per I.S. specifications.

MAIN DEFECTS IN LAMP MAKING

There are three main defects in lamp making on which most of the lamp makers concentrate their attention.

1. **The Carbon Fuzz or Dendrite.** Within the first three hours of burning of the lamp, there are long pointed black needle like formations on the mounting points of the lead-in-wires known as carbon fuzz. This is a misnomer because the black coloured carbon like substance is mainly tungsten and not carbon. Technically it is called dendrite.

The formation of carbon fuzz or dendrite is a clear indication of the presence of water vapour and impurities inside the lamp. The dendrite starts as a small black spot and slowly starts increasing in length and spreads over the lead-in-wire. Once this reaction takes place, it continues to grow—which means the water vapour has started helping the transportation

(*See Photo 9.1*)

cycle of tungsten which have come out from the filament but could not travel towards the shell wall due to pressure of the inert gas. THE CARBON-FUZZ LAMP HAS GOT TO BE DISCARDED.

Specially in auto-lamps where, the lamp is not properly aged, this defect is not easily detectable. The indication of a black-spot on electrode should act as a good warning to the shop floor supervisor. The lamp with carbon fuzz are the worst types and most unreliable. If not checked instantly and the cause eradicated, heavy losses will be incurred.

(*a*) Carbon fuzz generally occur when new argon-nitrogen circuit is made. The lamps from new circuit are to be heated in the exhaust bench for longer period. The procedure for proper evacuation, heating up and clearing the new circuit is mentioned earlier in the manufacturing chapter.

There is a general tendency of over gettering the lamps made from new circuit. The main cause of carbon fuzz is 'water vapour'. When this fuzz occurs, the operator puts new lamps every time and flushes them with the same gas mixture having water vapour. This causes heavy loss in production.

Actually one lamp only should be used and the other heads sealed or made non-operative—extra heating and a little shaking of the P_2O_5 towers will be necessary and a minimum of 3 or 4 passes made to wash the circuit.

Seasonal changes in gettering is to be made. The getter should be very finely ground and mixed with cryolite.

(*b*) Carbon fuzz occurs if the circuit is not kept worm during the night. Use 4 no. 100 watt lamps, so that the water molecules will remain in vapourised condition and will be driven off by the fresh gas when used for washing the circuit.

(*c*) When there is accumulation of yellowish substance and somewhat liquid material at the bottom of the tower, the carbon fuzz will be the lamp killer. The lamp coming out from this circuit will pass through all the rigours of test but after about 40% of it's life, it will start forming carbon fuzz.

So at the very first sign of discolourisation of P_2O_5 remake the purifying circuit.

(*d*) When the vacuum pump oil becomes old, it gets discoloured due to continuous passing of residual oxygen, water vapour and impurities inside the lamps, circuit and manifold connections. These pumps also leave vapours which get embeded on to the manifold and the heads. Finally the pressurised inert gas carries them inside the lamps which again cause carbon fuzz.

The pump oil should be periodically changed after flushing the pump, with new oil.

(*e*) After the day's work, the heads should not be left bare but covered with small test tubes and the hood cover put over them.

2. **Electrode Point Welding.** This is one of the commonest premature failing sign of the lamp. At the very first burning of the lamp, this occurs. The electrode hook melts down as a shining 'blob' and the filament end starts hanging loosely from the support. Sometimes both the ends of the filament are seen to be burnt off from the electrode hooks. The main reason for 'point-weld' is failure of vacuum or presence of oxygen inside the lamp.

If the defect persists on every batch and in quiet a good number a leak in the gas service line is indicated, whether it is being repeated from the same station or the same head-this has to be established at first.

(*a*) A faulty rubber base head will produce 'point-weld'.

(*b*) More joints on exhausting heads may cause the defect.

(*c*) Sometimes a loose hook-pinch on the filament also causes sparks and 'point-weld'.

The 'point-weld' defects are more pronounced in vacuum lamps. If the lamps are not properly heated, the lamps do not get properly evacuated of the impurities and oxygen gas. As told earlier, the vacuum gauge on the tester cannot indicate of the residual gases inside the lamp. So, proper heating of the lamp is most essential to make good vacuum lamps.

In gas-filled lamp, the gas inside lamp is nearly at the same pressure with outside atmosphere. In vacuum lamp the outside atmosphere is always trying to enter inside the lamp through the lead-in-wire joints and the 'exhaust seal'.

After completion of processing and carrying out all the tests, a vacuum lamp should not be cartonised the same day. It is advisable to keep the lamps bare and test them with the high frequency tester. Generally 2% to 3% lamps are found to show hazy colours meaning slow leaks through 'stem pinch'. These lamps surely will not burn for long time and at the end may end up with an 'weld point'.

Break these lamps carefully and check up very minutely the reason for 'stem pinch' leak. Mainly this happens due to bad selection of stem for mounting after annealing. In many cases the 'dumet joint' is found to be in exposed or semi-exposed condition.

When more and more lamps are found to have this defect of slow-leak—adjust the stem leads immediately.

It is to be remembered that the natural death of a lamp are decay of tungsten filament and undetectable slow leaks spreaded over months and years which is natural. With such a cheap item as electric lamp more perfection on hermetic sealing is not possible. At the same time, the hermetic sealing should not also give away within a month or two leave about one or two days. So, the 'stem slow leak' is a manufacturing fault which is to be accepted and means found to eradicate it.

Sometimes, in autolamps and in some high wattage lamps the stems are made of soda glass which suddenly die out with a spark. The heat produced inside an autolamp is not less and the heavy flow of current makes the stem pinch point more vulnerable to electrical discharge. Using 'lead glass' for stem is the only solution to avoid this 'point weld' problem.

3. **Overheated Lamps.** In electric lamps, more heat means more light. The conditions of this heating are not properly understood by many people in the manufacturing industry.

More heating means more heat made available at the filament and this heat must remain there constantly around the 'Langmuir sheath area'.

A defective 'lamp will' retain and utilise less heat at the filament and expell more heat through the outer shell of the lamp. These are known as 'hot lamps'.

This happens mostly due to the following faults in manufacturing which need be corrected and constant watch kept on their condition.

(*a*) Over or under filling of inert gas

(*b*) Over stretched mounting

(*c*) Improper light-centre

(*d*) Bad sealing of stem—stem bent to one side

(*e*) Using wrong getter

(*f*) Argon-nitrogen gas mixed with air.

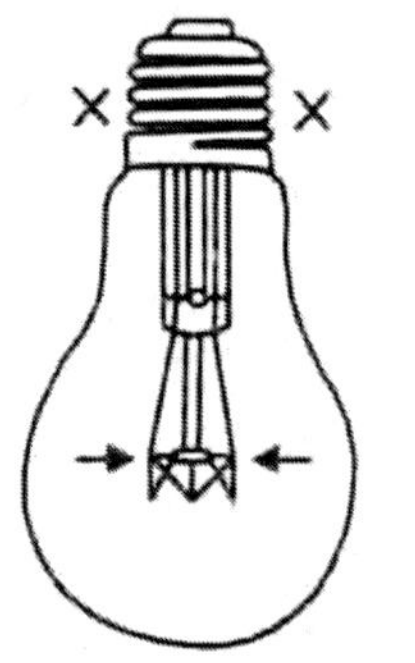

Fig. 9.1. → ← Heat concentration should be within this area. ×× No heat should concentrate here.

There are many reasons for getting this 'hot lamp' defect. A simple fault in raw materials selection, size, shape or faulty processing will lead to a hot lamp. A 'hot lamp, means a bad lamp'. This lamp can never pass through the quality tests. This cannot have the desired light and life.

To discourage this fault in lamp making, under advise from IEC, Indian Standards Institution has now adopted rise in the lamp cap temperature, as statutory standard.

On the shop floor the hot lamp is not easily recognisable. The experienced worker only recognises hot lamps which is really very unscientific way of working. Also it is not possible to test every individual lamp when high production figures are to be maintained.

The only way to avoid making 'hot lamp' is to bring perfection at every operation point. In mass production the general tendency is to blame the faulty operation of the machinery—but in actual practice the machine only repeats what has been set to it by it's master.

Lamp making is a scientific art work. Every person in this industry is to be made proud of his work—because he makes light, the gift of God.

INDIAN STANDARDS PUBLISHED

ON

TUNGSTEN FILAMENT LAMPS AND ACCESSORIES

IS : 418—1978. Tungsten filament general service electric lamps (*third revision*).

IS : 897—1966. Tungsten filament electric lamps for railway rolling stock (*first revision*).

IS : 1606—1979. Automobile lamps (*second revision*).

IS : 1885 (Part XVI/Sec. 3)—1967. Electrotechnical vocabulary : Part XVI Lighting, Section 3, Lamps and auxiliary apparatus.

IS : 1901—1978. Visual indicator lamps (*first revision*).

IS : 2261—1975. Lamps for flashlight (*first revision*).

IS : 2407—1963. Photometric integrators.

IS : 2592—1980. Lamps for lighting on board ships (*first revision*).

IS : 2596—1980. Bulbs (Lamps) for miners cap lamps (*first revision*).

IS : 3323—1980. Bi-pin lampholders for tubular fluorescent lamps (*first revision*).

IS : 6701—1972. Tungsten filament miscellaneous electric lamps.

IS : 7013—1981. Radio dial lamps (*first revision*).

IS : 7027—1973. Transistorized ballasts for fluorescent lamps.

IS : 8685—1977. Aircraft lamps.

IS : 8901—1978. Lamps for aerodrome lighting fittings.

IS : 8913—1978. Method of measurement of lamp cap temperature rise.

IS : 9206—1979. Dimensions of caps for tungsten filament general service electric lamps.

IS : 9589—1980. Electric Lamps for railway signalling.

Bibliography

Fundamentals of Light and Its Measurements

1. The Scientific Basis of Illuminating Engineering
   —*P. Moon*

2. Text Book of Illuminating Engineering
   —*J.W.T. Walsh*

3. Collected Works of Irving Langmuir

4. Lamps and Lighting
   —*H. Hewitt and A.S. Vause*

5. Modern Electric Lamp
   —*D.A. Clarke*

GLS and Halogen Lamps

1. Iodine Incandescent Lamps
   —*Tijen in Philips Tech. Review 23/59*

2. Lamps and Lighting
   —*Henderson and Marsden*

3. Badalex Catalogue for GLS Machine Group

4. Badalex Catalogue for H4 Halogen Hard Glass Group

5. Litex Catalogue for Halogen Lamp.